OPERATIONS TECHNOLOGY

Operations Technology

SYSTEMS AND EVOLUTION

ROBERT H. ROY

THE JOHNS HOPKINS UNIVERSITY PRESS
BALTIMORE AND LONDON

The Johns Hopkins University Press
701 West 40th Street
Baltimore, Maryland 21211
The Johns Hopkins Press Ltd., London

LIBRARY OF CONGRESS CATALOGING-IN-PUBLICATION DATA

Roy, Robert H.
  Operations technology.
  Includes index.
  1. System analysis. 2. Manufacturing processes. I. Title.
T57.6.R68 1986   658.4′03′2   86-7139
ISBN 0-8018-3340-X (alk. paper)

To My Dear Grandchildren
MEG
DAVE
MOLLIE
ELLEN
and to a Cherished Friend
WILSON SHAFFER

Contents

Illustrations

Preface

This book portrays the development of operations technology in a sequence that has been described as evolutionary. Whether or not that has been an original observation and whether or not it is true is now less important than my obligation to many friends and colleagues, first for genesis of the idea and then for helping toward its fulfillment.

A valued friend of long standing, Merritt A. Williamson, former Dean of the College of Engineering at Pennsylvania State University and later Ingram Distinguished Professor of Engineering Management at Vanderbilt University, became first Editor of *Engineering Management International*. From his editor's chair he invited me to submit a paper. Having but recently concluded service as Chairman of a Construction Investigation Committee for the city of Baltimore, an experience described herein, I did so. Later, Dr. Williamson invited a second paper. With encouragement from another friend and colleague, Carl Christ, Abram G. Hutzler Professor of Economics at Johns Hopkins University, I came to attempt this somewhat longer expository reach. Merritt Williamson, I am sorry to say, did not live to see this extension of his invitation. He died early in 1985.

For much general and specific assistance during preparation of the manuscript, I am greatly obliged to two other Johns Hopkins friends and colleagues, Willis C. Gore, Professor of Electrical Engineering and Computer Science, and Rodger D. Parker, Professor of Health Policy and Management in the School of Hygiene and Public Health and jointly Professor in the Mathematical Sciences.

For those portions of the book dealing with computers I am indebted not only to these colleagues but singularly and extensively to Elaine Rich, then Assistant Professor of Computer Science at the University of Texas at Austin, for her critiques of draft copy and, most of all, for her excellent book, a gift to me from Robert P. Rich of the Principal Professional Staff at the Applied Physics Laboratory, Chairman of Numerical Science and Lecturer in the School of Engineering, and Associate Professor of Biomedical Engineering at Johns Hopkins. Knowing this father and daughter has been a blessing.

Some of the case histories have been derived from my own professional experiences. Among the numerous others, perhaps the most unusual came from Frederic N. Smalkin, Esq., Magistrate of the United States District Court of Maryland, who supplied me with information hitherto unknown and guided me along a strange legal path. I thank him warmly for his interest. I also acknowledge my debt to Roszel C. Thomsen, former Chief Judge of this same District Court, for arranging the interviews.

In all of the case histories, "before" and "after" events were at times very elusive. Such was the situation for articulation until Howard P. Wampler, retired Vice-President of Waverly Press, Inc., where once we were fellow employees, supplied me with performance data that led me to Raymond D. Rodman, retired General Manager of the Hanover, Pennsylvania, plant of Doubleday & Company. Visits to Rodman's home were as pleasant as they were informative. I thank him and admire his remarkable accomplishments.

Serendipity played a considerable part in the accounts of glacé fruit processing and the manufacture of telephone cable. William S. Arnold, Jr., retired Department Chief, Engineering, of the Manufacturing Division of Western Electric Company, was the primary source of both cases, and Lorne R. Guild, retired Senior Staff Engineer at Western Electric, provided both contact and essential information.

My nephew, Roy Galloway, retired Vice-Chairman of Union Carbide Eastern, was in India and in charge of the battery plant there during the intrasystem automation described herein. His experiences during the turbulent time of Indian independence would make a better story than automation, but it would not be relevant here.

The research library account was familiar to me as a participant in the project, but the lion's share of credit for conception and execution belongs to Benjamin F. Courtright, then a graduate student at Johns Hopkins and later Professor and Chairman of Information Science at the University of Maryland at College Park. The torch that Courtright lit during his pioneering effort was fanned by John Berthel, then Librarian, and is today ably carried by Susan K. Martin, Director of the Milton S. Eisenhower Library. She and two members of her staff, David Miller and Marilyn Petroff, reviewed manuscript drafts and provided necessary information about current operations.

Another former graduate student, Donald L. Fink, Executive Vice-President DP/Associates, Inc., made it possible for me to meet and interview George Goodman, Vice-President, T. Rowe Price Associates, Inc., about the programmed operation of the money market fund. I was for a time fearful that proprietary information could be inhibiting. On the contrary, I was allowed complete freedom, and I thank both of these gentlemen for the privilege.

The account of robots would not have been possible without the very considerable help received from Rollie Woodcock, Corporate Communications Manager of Intelledex. He supplied me not only with suitable technical papers but also with necessary illustrations—all accompanied by helpful telephone calls from Oregon. I am very grateful to him and hope that some day we may meet.

The description of use of computers in medical diagnosis is based on papers supplied to me by Jonathan M. Links, Assistant Professor of Environmental Health Science in the School of Hygiene and Public Health, and, jointly, in Nuclear Medicine in the School of Medicine, at Johns Hopkins. Dr. Links became known to me through a mutual friend, Kathleen Prendergast, Research Project Coordinator in the Division of Nuclear Medicine and Radiation Health Science at the School of Hygiene and Public Health.

For her excellent drawings I am very much obliged to Dean L. Pendleton of the Illustration Division at Johns Hopkins.

Proofreading, always an arduous task, was made pleasant by expert help from my niece, Nancy Manger. Her companionship was as much appreciated as her vigilance.

The Johns Hopkins University Press is again my publisher. With them, I have been made to feel at home by Anders Richter, Editorial Director. He has been frank, knowledgeable, supportive, and always friendly. To others at the Press thanks are also due: to Barbara Lamb, Nancy Essig, and James Johnston, and especially to Penny Moudrianakis, for her careful copyediting.

Lastly, for an interesting and useful word, *commensalism*, I am obliged to Sigmund S. Suskind, University Professor of Biology at Johns Hopkins.

Without these many contributions, this book would not exist today. To say that, however, is not enough. Every assistance has been provided with graciousness beyond my power to thank.

Robert H. Roy

OPERATIONS TECHNOLOGY

ONE

Introduction

Whenever the goods or services of any organization are provided to customers or clients by means of a succession of related operations or processes, some system of operations technology exists. Such systems are as diverse as they are numerous, but their technologies can be categorized and their evolutionary development traced. Metaphorically speaking, these evolutionary paths describe an almost closed circle.

All of the operational system technologies to be discussed, save perhaps the last, still exist; as yet, none has become extinct. Like related species in the living world, system technologies coexist, often in hybrid combinations. Their evolution has been about as follows:

Jobbing systems
Articulated systems
Balanced systems
Continuous systems
Automated systems
Programmed systems
"Intelligent" systems

For each of these technologies, discussion will focus upon time intervals for elements that can be combined to depict either whole systems or component parts of them. All operational systems of whatever kind can be represented by suitable aggregations of the following elements:

Moving unit or batch of product or service to a work center

This element measures the time required to move the product or service (hereafter called the "job") into or out of the system, as well

1

as between work centers.[1] When jobs are large, bulky, or immovable, as in the case of building construction, work and workers may move to and about the job site. In either case, time intervals for moves, denoted by the symbol *M*, may be very short or very long, but, as with all time intervals, must be positive.

Processing job at work center

Processing is the primary purpose of any system and usually dominates system cost. Time elements for processing, denoted by the symbol *P*, measure intervals required to perform work upon jobs at work centers. Processing intervals may be long or short, but must be positive.

Stoppages

Stoppages are intervals during which systems, or parts of systems, cease operations, from either external or internal causes.

Stoppages characterized as *external* are caused by diminution or absence of demand (loss of market), by lack of materials or supplies, or by the concerted absence of workers, should they elect to strike. External stoppages are important and costly to all system technologies, but since they are unpredictable and imponderable, will not be considered in the comparisons to follow.

Stoppages characterized as *internal* are caused by breakdowns, errors, and, on occasion, by severe imbalances. Such stoppages are much less significant in jobbing systems than in successor forms. They are to a degree controllable by provision of stand-by labor and equipment, preventive maintenance, fail-safe controls, and redundant design and circuitry, all of which involve trade-offs between the cost of system stoppages and the costs associated with measures to prevent interruptions to system operation.

Stoppage intervals from internal causes will be identified by the symbol *H* (for "holding time"). In the equations that follow, *H* will not appear ($H = 0$), on the assumption that system measures apply to performance without stoppages. Where appropriate in system comparisons, holding times and their effects will be introduced.

---

1. A *work center* is a facility at which jobs are processed. A work center may consist of a single operator or machine or both. A work center may also comprise multiple *work stations*, each having common operator, equipment, and cost characteristics.

Delays to work in process

When a job has been moved to a succeeding work center and there joins a queue of other work in process, the job will be delayed until preceding jobs have been processed, but the processing function at that center will not be interrupted. The disutility of inventory-in-process delay derives from impaired service to customers and from the carrying cost of inventory (working capital and floor space). Time intervals for inventory delay, denoted by the symbol D, can be zero but also can be very large.

Sequencing delays

Sequencing delays are the converse of inventory delays: they are delays to work stations rather than to in-process inventory. Such intervals of idleness are incurred when a work center completes a process *before* the next job is moved to that center. Sequencing idleness may also be incurred when a work station is stopped for want of attention from an operator otherwise engaged.[2] These classes of delay, denoted by the symbol I, often involve operating as well as capital costs. Time intervals for I must be positive, but can be zero. If I is positive, the corresponding value for D must be zero, and vice versa.

All five categories of elemental time intervals can be illustrated by regarding a traffic light as a single work center, controlling flow in north-south and east-west directions.

---

2. Sequencing delays of this kind were the focus of studies of circuit assignments to telephone operators in the days when all calls were operator assisted. Depending upon subscriber demands and the service ("attention time") required of the operator, there were occasions when subscribers would have to wait for the inquiry, "Number, please," while the operator serviced other calls. These intervals of customer waiting were called "interference time." Assignment of an excessive number of circuits would decrease wage costs but would increase to perhaps intolerable levels the interference "costs" borne by subscribers, which shows, by the way, that such costs are sometimes externalized.

Interference delays are present in all multiple-machine operations (looms, automatic screw machines, Monotype casting machines, etc.) where there are more machines than operators. Minimizing costs in such circumstances requires weighing the conflicting values of machine and operator assignments against the frequencies and attention times of stoppages and the frequencies and durations of interference time. See W. G. Duvall, "Machine Interference," *Mechanical Engineering*, 58, no. 8 (1936): 510–14.

As long as cars are passing through the intersection in any one of the four directions, the center may be said to be providing service, measured by the interval P, and measured for the passage of each car by the moving interval M.

When cars are moving through the intersection during a green interval, there are likely to be cars waiting in the red direction. These waiting cars are undergoing delay intervals D.

When cars are delayed at a red light, and *no* cars are passing in the green direction, the system is not providing service but is undergoing a sequencing delay I. The waiting cars at the red light are experiencing "interference."

If no cars are seeking passage through the intersection in either direction, the system is not providing any service, a stoppage created by an absence of "customers." Prevalence of such a condition would suggest that the capital cost of providing the light has been wasted.

If, on the other hand, a plethora of cars seeks to pass the intersection in both directions at the same time, the system itself could be stopped by "gridlock," an internal source of stoppage denoted by H. Should these intervals be sufficiently frequent, conversion of the intersection into a two-level "cloverleaf" could be contemplated. This could reduce or eliminate holding time H, but would require capital for construction.

Improvements in system performance have been realized by reductions in all of these categories.

Moving times have been reduced by technological advances in materials handling, in the storage, retrieval, and transmission of information, and by evolutionary developments in system technology.

Processing intervals at work centers have been reduced by a long succession of innovations: new technology, substitution of capital for labor, mechanization, standardization, interchangeability, work simplification and measurement, human factors, numerical controls, participative management, automation, and so on.

Delays and idle intervals have been reduced dramatically by system mutations. In transitions from jobbing systems to articulated, balanced, and continuous systems, very large reductions in delay times have been realized by sacrificing capabilities for variety in exchange for speed. In later mutations to automated and programmed systems, capabilities for variety have been regained, without sacrifices in delay times and costs.

Stoppages from internal causes have been reduced by providing stand-by labor and equipment, fail-safe controls, test programs, redundant design and circuitry, random inspection, and the like.

If and when artificial intelligence creates what I have called "intelligent" systems, the metaphorical circle will be closed: we shall have operational system technologies not only capable of variety with speed but also possessing the humanlike ability to adapt, to program and operate upon unforeseen contingencies.

COMPARATIVE SYSTEM CHARACTERISTICS

In succeeding chapters the characteristics of each of the seven operational system technologies named will be described by examining the time dimensions of the five elemental intervals. The relationships deriving from these configurations will be compared and modeled for

> Total time
> Cycle time
> Operational efficiency
> System efficiency
> Working-capital requirements
> Work-in-process inventory
> Operational floor-space requirements

Discussion will be confined to events taking place between points of entry into a system and completion of the product or service the system provides. Raw materials, fixed capital, maintenance, personnel, energy, and other matters germane to organization and operation will be used in support of discussion as deemed necessary, but events between commencement and completion will be the central theme.

NOTATION

Examples showing how the following notation applies are given in Fig. 1.1 and in succeeding chapters.
Let

$1, 2, \ldots, i$ = Superscripts, each of which identifies a particular job or unit or batch of product or service. Ordinal progression is not required.

$1, 2, \ldots, j$ = Subscripts, each of which identifies a single work center. Subscripts followed by single asterisks (*)

identify work centers through the longest path of a system. Ordinal progression is not required.

M_j^i = Moving time for the ith job to a succeeding work center, or into or out of a system.

$_jD^i$ = Delay time for the ith job *before* processing at the jth work center.

D_j^i = Delay time for the ith job *after* processing at the jth work center.

P_j^i = Processing time for the ith job at the jth work center.

$_jI^i$ = Idle time before processing at the jth work center, awaiting completion of the ith job at the preceding work center.

H = Stoppage intervals attributable to internal causes.

T^i = Total time for the ith job to progress through the system.

C^{i*} = "Cycle time" for the ith job to progress through the system's longest path.

W = Working-capital (in contrast to fixed-capital) requirements for the operation of a system, expressed, for purposes of comparison, in units of time.

V = Inventory of work in process, expressed in units of time.

A = Floor-space requirements for storage of work in process, expressed in units of time.

S = System output in units of "standard time."[3]

t = A moment in time. Example: the end of an accounting period, e.g., midnight, December 31.

Δt = An interval of time. Example: an accounting period, e.g., a month, quarter, or year.

$(E_p)_{\Delta t}$ = Operational efficiency of a system over time Δt.

$(E_s)_{\Delta t}$ = System efficiency over time Δt.

$(\ )^{**}$ = A restriction that limits times to elements that

---

3. Standard times for processing products or services are usually set by time study and serve the same useful purpose as the concept of par in golf: to provide a consistent measure of performance. Golfers' performances are measured and compared by strokes above and below par; operators' or machines' performances are measured and compared by relating output in standard minutes to input in actual minutes taken. In the absence of standard times as measures of output, "productive" times may be taken from time cards, or physical units of product or service may be used.

have been completed by time t or during the
interval Δt.

$\overset{i}{\Sigma}$ = Subject to the single and double asterisk
restrictions, only elemental times for all jobs in
process at time t, or within interval Δt, are to be
included.

$\overset{j}{\Sigma}$ = Subject to these same restrictions, elemental time
values at all work centers utilized are to be
included.

To represent a system interoperation, one can arrange events
either serially, with one operation followed by another and another,
or in parallel, with certain operations performed concurrently. There
can be as many operations and as many channels as necessary. An
example of a three-channel sequence, its longest path, and of the
notation is shown in Fig. 1.1.

SYSTEM CHARACTERISTICS

Total time

Total time for passage of the ith job through a system is the sum of
all elemental times necessary to complete that job:

$$T^i = \overset{j}{\Sigma} (M^i_j + {}_jD^i + P^i_j + D^i_j). \tag{1.1}$$

Cycle time

Cycle time for any single job traversing any system is the interval
between commencement of work and completion of a unit or batch
of product or service. Cycle time for progression of the ith job, C^{i*},
is measured by the sum of the longest-path move times, delay times,
and processing times:

$$C^{i*} = \overset{j}{\Sigma} (M^i_{j*} + {}_{j*}D^i + P^i_{j*} + D^i_{j*}). \tag{1.2}$$

Idle time, ${}_jI^i$, has not been included in equations (1.1) and (1.2),
because work-center idleness caused by sequencing does not increase either total time or cycle time. If P_j is idle, that center will
be ready to begin processing as soon as work has been moved to it;
hence ${}_jD^i$ will be zero.

Fig. 1.1   Flow process chart for a single job processed through a three-channel system. Interval elements are shown in series (each single channel) and in parallel (all three channels). Plus signs indicate positive time intervals. *H* has been assumed to be zero.

M_1 +		M_8 +		M_{10} +	
$_1D$ +	$(_1I = 0)$	$_8D$ +	$(_8I = 0)$	$_{10}D$ +	$(_{10}I = 0)$
P_1 +		P_8 +		P_{10} +	
D_1 +		D_8 +		D_{10} +	
M_2 +		M_9 +		M_9 +	
$_2I$ +	$(_2D = 0)$	$_9D$ +	$(_9I = 0)$		
P_2 +		P_9 +			
D_2 +		D_9 +			
M_3 +		M_3 +			

$_3D$ + $(_3I = 0)$

Total time (eq. [1.1]) is the sum of the time intervals in all three channels:

$$T = M_1 + {}_1D + P_1 + D_1 + M_2 + P_2 + D_2$$
$$+ M_3 + {}_3D + P_3 + D_3 + M_4 + {}_4D + P_4$$
$$+ D_4 + M_5 + P_5 + D_5 + M_6 + {}_6D + P_6$$
$$+ D_6 + M_7 + M_8 + {}_8D + P_8 + D_8 + M_9$$
$$+ {}_9D + P_9 + D_9 + M_3 + M_{10} + {}_{10}D + P_{10}$$
$$+ D_{10} + M_9.$$

Cycle time (eq. [1.2]) is the sum of only those intervals that lie along the longest path, indicated by the bold line:

$$C^* = M_{10} + {}_{10}D + P_{10} + D_{10} + M_9 + {}_9D$$
$$+ P_9 + D_9 + M_3 + {}_3D + P_3 + D_3 + M_4$$
$$+ {}_4D + P_4 + D_4 + M_5 + P_5 + D_5 + M_6$$
$$+ {}_6D + P_6 + D_6 + M_7.$$

NOTE: $_2I$ and $_5I$ are shown in the illustration but are not included in the summations, because these intervals represent delays to the *system* rather than to the *job*.

(Left channel continued, with bold line path:)

P_3 +
D_3 +
M_4 +
$_4D$ + $(_4I = 0)$
P_4 +
D_4 +
M_5 +
$_5I$ + $(_5D = 0)$
P_5 +
D_5 +
M_6 +
$_6D$ + $(_6I = 0)$
P_6 +
D_6 +
M_7 +

Operational efficiency

Conventionally, operational efficiencies are calculated by summing measures of output over a suitable time span—a 13-week accounting quarter, for example—and dividing by the sum of processing times required to produce that output. For present purposes, to better compare system technologies, the denominator will include moving and idle intervals, the rationale being that, during moving, work is done on the product or service, and work-center expense continues to accumulate during idle intervals. In the following expression for operational efficiency, it is assumed that the numerator is in units of standard time:

$$(E_p)^{**}_{\Delta t} = \frac{\overset{i}{\sum}\,\overset{j}{\sum}\,(S^i_j)^{**}_{\Delta t}}{\overset{i}{\sum}\,\overset{j}{\sum}\,(M^i_j + {}_jI^i + P^i_j)^{**}_{\Delta t}}. \tag{1.3}$$

System efficiency

As above, within comparable time spans, the sums of all moving and delay times, and work-center and idle times, whether parallel or serial, provide measures of system inputs. When divided into compatible measures of output, as above, the quotient is a measure of *system* efficiency:

$$(E_s)^{**}_{\Delta t} = \frac{\overset{i}{\sum}\,\overset{j}{\sum}\,(S)^{**}_{\Delta t}}{\overset{i}{\sum}\,\overset{j}{\sum}\,(M^i_j + {}_jD^i + {}_jI^i + P^i_j + D^i_j)^{**}_{\Delta t}}. \tag{1.4}$$

Several observations can be made about these ratios:

1. Because of inevitable variances in measurements, the relative accuracy of each efficiency ratio between one time period and another is much greater than the absolute accuracy of any single quotient. Computation therefore serves best for purposes of comparison and control, for which purposes efficiency ratios can be very useful.

2. In the mutations of system evolution, intensive and successful efforts have been made to increase outputs relative to inputs. Operational efficiencies have played an important part in stimulating these efforts.

3. Very large improvements in system efficiency have been achieved as well, not only through improvements in processing, but also in the diminution of delays and idle time. But use of system efficiency ratios has been conspicuously lacking, especially where most needed: in jobbing systems, the subject of the next chapter.

Working-capital requirements

That portion of working-capital requirements not diminished by payments on account or in advance, or increased by extensions of credit, will be proportional to the interval between resource input and completion of the good or service provided by the system. For the ith job, that interval can be represented by the sum of move times, delay times, work-center times, and idle times:

$$W^i \sim \overset{j}{\Sigma} \, (M^i_j + {}_jD^i + {}_jI^i + P^i_j + D^i_j). \qquad (1.5)$$

Idle time I is included because it involves resource input during idleness intervals at essentially the same higher cost per unit time as processing intervals P.

Work-in-process inventory

If the ith job, as postulated above, had not been completed, its contribution to in-process inventory would be proportional to the same summation shown for T in equation (1.1), restricted only to those elements that had been completed at time t:

$$(V^i)_t \sim \overset{j}{\Sigma} \, (M^i_j + {}_jD^i + P^i_j + D^i_j)_t^{**}. \qquad (1.6)$$

Operational floor-space requirements

Space requirements for work in process are of little consequence in some systems—for example, where the product is jewelry or pieces of paper—but are critical in others, where products may be heavy or bulky. However small or large the space requirements for a product or service may be, in-process storage area or volume will depend not only upon size but also upon the time interval during which work is progressing through the system. Therefore, for any kind of product or service, space for the ith job is proportional to the same time intervals summed for the work-in-process inventory:

$$(A^i)_t \sim (V^i)_t \sim \overset{j}{\Sigma} \, (M^i_j + {}_jD^i + P^i_j + D^i_j)_t^{**}. \qquad (1.7)$$

For comparisons between and among operational system technologies these models for a single unit or batch of product or service may suffice, but convention may be better served by including all jobs in process at time t. The proportions then become

$$(W)_t \sim \overset{i}{\Sigma} \overset{j}{\Sigma} \, (M^i_j + {}_jD^i + {}_jI^i + P^i_j + D^i_j)_t^{**} \qquad (1.8)$$

and

$$(V)_t \sim (A)_t \sim \overset{i}{\Sigma} \overset{j}{\Sigma} \, (M_j^i + {}_jD^i + P_j^i + D_j^i)_t^{**}. \tag{1.9}$$

These time intervals do reflect magnitudes of working capital, in-process inventory, and floor space, but they are not commensurate with the monetary units customarily used to measure the first two nor with the area or volume units used to measure the third. For working capital and inventory, conversion from time units can be made by multiplying work-center, idle, and moving times by the average direct cost per unit of time, and all the elemental sums by an interest rate representing the time-dependent cost of money.

In most cases the unit cost of processing, idleness, and moving, derived from wage and equipment costs, will be much greater than the rate used to calculate the cost of money. In some instances, however, cycle time is made inordinately long by excessive delays, and holding time may be diminished only by increasing direct time. It then becomes desirable to make interval decrements and increments commensurable. This may be done as follows:

If C = the average cost per unit of time for processing, idleness, and moving, R = the annual interest rate, and Δr is understood to be not a single value but the time interval applicable to each job for the calculation of interest, we can convert units of time to commensurable monetary units by

$$(\$W)_t \sim (V)_t \sim \overset{i}{\Sigma} \overset{j}{\Sigma} \, \{ \, C \, (M_j^i + P_j^i + {}_jI^i)$$
$$+ R[C(M_j^i + P_j^i + {}_jI^i) + {}_jD^i + D_j^i]_{\Delta r} \}_t^{**}. \tag{1.10}$$

No attempt will be made to develop a similar model relating space costs to time for passage of a good or service through a system.

In all the preceding equations, we have assumed, as said before, that system performance was not interrupted by stoppages, from either external or internal causes. Thus, for internal causes, $H = 0$. Should H be positive, such intervals would increase the summations in equations (1.1), (1.2), (1.5), (1.6), (1.7), (1.8), (1.9), and (1.10), and the denominators of equations (1.3) and (1.4).

Operational systems range in size and complexity from very small, single-person establishments that in evolutionary terms are likely to remain primeval, to enormous, multinational operations that have evolved in sophisticated ways and are much too large for textual representation. Whether large or small, most operational systems are very likely to embody "hybrid" rather than "pure" technological characteristics—for example, jobbing sequences, articulated arrangements, some automated operations, and, increasingly, system components that have been programmed.

To make relevant comparisons between operational system technologies, it is therefore necessary to simplify by postulating sequences of modest size, technologies operating as "pure" jobbing systems, "pure" articulated systems, "pure" balanced systems, and so on.

This will be done first by assuming a five-operation, single-channel sequence and assigning to its components numbers that will, with reasonable fidelity, reflect the behavior of several of the systems under consideration.

About the models to be used in making comparisons there is an implication of complexity that may need dispelling, an implication that substitution of the numbers called for by the elemental components would require a great deal of additional record keeping. This is more than a half-truth, yet, as will be shown, some measurements not conventionally recorded or analyzed—for example, delays and cycle times—might be made with profit. Other items, such as processing times and costs, work-in-process inventories, working-capital requirements and related monetary costs, are the results of customary accounting routines.

Given the capabilities of present-day computers, data taking, if only for the occasional purposes of system analysis, could yield significant benefits in improving and controlling the performance of operational systems—an assertion that will be reinforced by examples.

TWO

Jobbing Systems

Least satisfactory of the eight adjectives used in the Introduction to label system technologies is *jobbing*. Derived from the term *job printer*, which engages in printing anything from a few business cards to thousands of multipage, illustrated catalogs, *jobbing* properly identifies customer- or client-demanded tasks as the primary purposes of jobbing systems, but it does not adequately connote the prevalence or diversity of such systems throughout the operational world.

A print shop is a jobbing system, to be sure; so also in comparably recognizable ways are general-purpose machine shops and almost all repair, remodeling, and service facilities. Less well recognized as jobbing system technologies, however, are many other kinds of customer- and client-oriented systems: outpatient and inpatient services provided by hospitals; legal and judicial processes; custom manufacturing, including construction and other one-of-a-kind projects; police and fire departments; innumerable bureaucratic operations in government; equally numerous office procedures in the private sector; and much of advertising and merchandising.

JOBBING SYSTEM CHARACTERISTICS

The single word that best characterizes *every* jobbing system is *variety*. Within chosen specialties—printing, health care, repair, advertising—each such system must be equipped and staffed to provide the diverse goods and services sought by customers and clients. For all such systems, variety is the name of the game, the *raison d'être* for jobbing system technologies.

Frequent but less-universal are other advantageous characteristics of jobbing systems:

Relative to successor technologies, jobbing systems are more easily conceived, created, and expanded or contracted. Unlike their evolutionary progeny, all of which require intensive and extensive preliminary analysis, jobbing systems need but little "prenatal design." It is easy to "go into business."

Related to this start-up simplicity are two other advantages: jobbing systems usually require less fixed capital than successor systems, and utilize general-purpose machinery, equipment, and personnel, grouped according to the processes performed rather than the products or services to be provided. There are degrees of difference, but jobbing systems tend to be labor rather than capital intensive.

Once conceived, funded, equipped, and staffed, jobbing systems confront corollary characteristics that complicate and constrain their operations:

Demands for service arrive at points of entry at times determined by customers and clients, or by such unforeseen contingencies as accidents or illness.

Demands for service at points of entry may be accepted, deferred, or declined by many jobbing system managers, but in some jobbing systems, acceptance of demand is obligatory. The job printer can turn away or defer a proferred order, but police and firemen must respond to calls for help, and those presiding over hospital emergency rooms would suffer opprobrium if seriously injured patients were turned away.

Being accidentally or volitionally determined, arrivals are variously distributed over time; there are peaks and valleys in demand, sometimes predictable and sometimes not. Unless systems are overstaffed or overequipped, as is sometimes necessary, peaks cause system overload and delayed service; valleys, caused by slack demand and perchance aggravated by overstaffing and overequipping, threaten loss of revenue and costly system idleness.

Jobbing systems are characterized by multiple work centers and extensive division of labor. One work center or worker may repeatedly perform Operation 1, another Operation 2, a third Operation 3, and so on. Depending upon system or project size, there are likely to be many work centers: hundreds of divided labor operations have been used in the manufacture of this book; thousands of divided labor sequences are to be found in the bureaus of local, state, and national government.

Multiple work centers and extensive division of labor are to be found in other system technologies as well, but as will be shown, the separation of work into component parts exacts penalties that are unique to jobbing systems.

Routes followed by goods and services progressing through jobbing systems are neither uniform nor consistent. One product or service may follow a sequence of, say, 20 operations in a kind of numerical order (P_{1-20}); another may follow the same order but use only 15 components; a third may need only 4 component operations (P_{1-4}), while 5 others (P_{5-9}) are performed concurrently; and a fourth may call for several backtracking repetitions of the same operation $(P_{1-2-1-3-4-1})$.

Inconsistencies in work-center sequences are more than matched by heterogeneous time requirements: differences between one work center and another, and differences between one task and another at the same work center. Processing intervals can be and sometimes are the same, but inequalities are the rule, and the magnitudes of work-center and job differences are often large.

Prevalence of job and operation inequalities, wherein $P_1^1 \neq P_1^2 \neq P_1^3 \neq P_2^1 \neq P_3^1 \neq \ldots P_{ji}^i$, makes backlogs of work mandatory at jobbing system work centers. To maintain operational stability, work-center in-process inventories must be large enough to cushion whatever the magnitudes of these variations may be.

As a consequence of in-process inventories large enough to prevent work-center idleness, each job must await processing at each successive work center by an average interval equal to the processing time for all preceding jobs in the work-center queue. Waiting time for the ith job at the jth work center will be

$$_jD^i = \sum^i (P_j^1 + P_j^2 + P_j^3 + \ldots P_j^{i-1}).$$

If a first come–first served rule is violated by giving first priority to a new arrival, waiting time for all the jobs displaced will be increased by the interval required to process the job given preference.

Given the magnitudes of delay times $_iD^i$ and D^i_j and the fact that in-process and interoperational floor space are directly related to these durations (equation [1.7]), jobbing system values for V and A, relative to other system technologies, are very large.

Combinations of multiple work centers and extensive division of labor, lengthy delays, and large in-process inventories inevitably lead to long time intervals between commencement of work and emergence of completed goods and services. Total time, T, and cycle time, C^*, are much greater for jobbing system technologies than for successor systems (equations [1.1] and [1.2]).

Insufficient or inadequate control of jobbing systems, at points of entry and during processing, make late deliveries commonplace and "lost" items frequent.

As calculated in equation (1.3), because of inclusion of moving and idle times in denominators, and because values of P are likely to be higher in jobbing processes, operational efficiencies, E_p, are lower than comparable ratios in successor systems.

Addition of the considerable delays that characterize jobbing systems to denominators (equation [1.4]) inevitably makes system efficiencies, E_s, very much lower than comparable ratios for successor systems.

Low operational and system efficiencies are directly related to long intervals between commencement and completion of jobs, and also to a different, subjective criterion: system *effectiveness*. Does the system do well in the estimation of its customers and clients? Does it perform its intended tasks in a satisfactory way? In this context there are more than a few effective jobbing systems, but in general, jobbing systems are likely to be less effective than successor forms.

Most, though not all, of these constraining characteristics derive from the variety of goods and services that jobbing systems provide. Processing times may vary over wide ranges in the manner shown in Table 2.1, in which units of time have been assumed for 20 consecutive jobs processed at five work centers. This very simplified single-channel representation shows no variation in work-center se-

Table 2.1 Assumed times in minutes for processing 20 consecutive jobs at five work centers

Job	Work center				
	P_1	P_2	P_3	P_4	P_5
P^1	5	20	12	40	17
P^2	56	218	38	111	85
P^3	105	57	23	308	216
P^4	5	20	12	40	17
P^5	9	33	0	63	8
P^6	73	82	54	417	321
P^7	5	20	12	40	17
P^8	5	20	12	40	17
P^9	143	49	263	308	45
P^{10}	6	13	20	5	0
P^{11}	5	20	12	40	17
P^{12}	441	68	71	173	99
P^{13}	5	20	12	40	17
P^{14}	81	67	142	163	218
P^{15}	21	38	375	10	87
P^{16}	5	20	12	40	17
P^{17}	38	409	138	193	305
P^{18}	61	203	192	78	176
P^{19}	5	20	12	40	17
P^{20}	17	63	82	118	71

quences, nor any backtracking. There are only a few work centers; therefore there is little division of labor, and problems of jobbing system control have not been made evident. However, the assumed time intervals, whether in hours, minutes, or days, do permit reasonably accurate comparisons between and among system technologies.

For reasons that will be made clear in Chapter 6, intervals shown for the first unit or batch of product or service ($P_1^1 = 5$, $P_2^1 = 20$, $P_3^1 = 12$, $P_4^1 = 40$, $P_5^1 = 17$) have been used to evaluate characteristic jobbing system behavior. (These values, again for reasons that will be made clear later, are repeated for seven other jobs in the table: P^4, P^7, P^8, P^{11}, P^{13}, P^{16}, and P^{19}.)

Additional elemental times for moves and delays are assumed in Table 2.2, where the sum of moves and delays is shown to be about 90 percent of total time, leaving only 10 percent for actual processing of the product or service. This degree of imbalance is by

Table 2.2 Element intervals for Job No. 1

Element		Time (in min.)	Operational step
M_1^1	=	6	Move Job No. 1 to Work Center 1.
$_1D^1$	=	180	Await processing.
P_1^1	=	5	Process at work Center 1.
D_1^1	=	20	Await move to Work Center 2.
M_2^1	=	6	Move to Work Center 2.
$_2D^1$	=	92	Await processing.
P_2^1	=	20	Process at Work Center 2.
D_2^1	=	20	Await move to Work Center 3.
M_3^1	=	8	Move to Work Center 3.
$_3D^1$	=	209	Await processing.
P_3^1	=	12	Process at Work Center 3.
D_3^1	=	20	Await move to Work Center 4.
M_4^1	=	3	Move to Work Center 4.
$_4D^1$	=	104	Await processing.
P_4^1	=	40	Process at Work Center 4.
D_4^1	=	20	Await move to Work Center 5.
M_5^1	=	9	Move to Work Center 5.
$_5D^1$	=	101	Await processing.
P_5^1	=	17	Process at Work Center 5.
D_5^1	=	20	Await move-out.
M_6^1	=	4	Move out of system.

Totals

$$\sum^i M_j^1 \ = \ 36$$

$$\sum^i {}_jD^1 \ = \ 686$$

$$\sum^i P_j^1 \ = \ 94$$

$$\sum^i D_j^1 \ = \ 100$$

no means exaggerated for jobbing system technologies. Actual data from various sources conform with the assumed proportion and so does rationalization: large preoperational in-process inventories provide managerial and employee security and system stability; imprudent acceptance of new obligations under conditions of system overload is more likely than prudent declination; and delays are aggravated by human propensities for setting aside that which is difficult, and instead, deciding to do that which is easy.

In the flow chart all postoperation delays have been assigned 20 units of time, on the assumption that products (or documents) are conveyed into, through, and out of the system by move men (or interior mail clerks) making rounds at scheduled intervals. Moving times have been assigned short intervals on the assumption that work centers, although not juxtaposed, are not far apart.

Preoperation delays have been arbitrarily chosen to reflect the varieties and magnitudes of the work-in-process inventories that characterize jobbing systems.

Based upon these assumed data, previously defined characteristics can be calculated for purposes of system comparisons. All time units are assumed to be in minutes.

Total time and cycle time

Since there is but one channel in the assumed sequence, total time T^1 and cycle time C^{1*} (equations [1.1] and [1.2]) are the same:

$$T^1 = C^{1*} = 36 + 686 + 94 + 100 = 916 \text{ minutes.}$$

Operational efficiency

Data for system inputs and outputs have not been assumed for a time interval, Δt, as is customary, so it is possible to calculate E_p (equation [1.3]) only for the single job chosen, for comparison with other system technologies. For this purpose, assuming that system output S_j^1 is set at 94 standard minutes for Job No. 1, the following ratio will serve the purpose of system comparisons:

$$E_p^1 = 94/(36 + 0 + 94) = 94/130 = 0.723 = 72.3\%.$$

System efficiency

Based upon the same assumptions, system efficiency E_s^1 (equation [1.4]) in the jobbing mode will be

$$E_s^1 = 94/(36 + 686 + 0 + 94 + 100) = 94/916$$

$$= 0.103 = 10.3\%.$$

Working-capital requirements

As defined in equation (1.5), working capital is proportional to the time span over which resource inputs are required for operational purposes. Since in this case idle time $_jI^1$ has been assumed to be zero, operational working capital requirements will be proportional to T^1 and C^{1*}:

$$W_j^1 \sim T_j^1 \sim C_j^{1*} \sim 36 + 686 + 0 + 94 + 100 = 916 \text{ minutes.}$$

Inventory of work in process

In-process inventory is expressed by the proportionality shown in equation (1.6). At the moment of completion of Job No. 1, this proportionality is the same as that for working capital if, as has been assumed, the value of $_jI^1$ is zero.

Floor-space requirements

Floor-space requirements for Job No. 1 depend upon the area needed for that particular job. This may be assumed to be constant, but when Job No. 1 is moved to the next work station, space must be provided for it, and that space will not be available for any other work while Job No. 1 is awaiting processing or moving. Time is therefore of the essence, so to speak, and these criteria may therefore be expressed as above—that is, as proportional to 916 minutes.

All of the foregoing summations have been applied to a single job. Equations (1.8), (1.9), and (1.10) more realistically apply to whole-system status for partially completed work at time t, but, unfortunately, real data are not available, nor am I capable of making suitable assumptions. Comparisons with other system technologies will therefore be made by using equations (1.1) to (1.7) for a single job.

THREE

Improvements in
Jobbing System Operations

The best way to alleviate the many constraints that attend jobbing system operations is to change the technology to one or another successor form. This, however, is not always possible. Many jobbing systems are not amenable to technological conversion. But this does not mean that operational improvements are beyond reach.

Among the constraints listed in Chapter 2 are two that have not yet been illuminated by equations and proportionalities:

1. Excessive division of labor and, not infrequently, unnecessary operations, exact penalties—disutilities—that are unique to jobbing system technologies.

2. Unless a first come–first served, or some other fixed-sequence, rule is adopted and enforced, to establish the order of work at each center, control of jobbing systems must be exercised, not just at points of entry, but at each work center.

These constraints and the means for dealing with them are examined here and will be exemplified in the case histories of Chapters 4 and 5.

DIVISION OF LABOR

In Table 2.1, Job P^5 shows a processing interval of zero at Work Center P_3. While this is intended to mean that this particular job did not require service at this work center, it may be interpreted otherwise, to mean that *all* operations at Work Center P_3 might be eliminated. By either interpretation, Job P^5 did not have to move to Work Center P_3, did not have to await processing there nor require processing time, did not have to await the move to Work Center P_4, and did not require the time to move the job to that center. Thus,

21

elimination of P_3 has saved not only the processing time there but the intervals for M_3^5, $_3D^5$, D_3^5, and M_4^5 as well.

Improving jobbing system performance by eliminating unnecessary operations is not a fanciful prescription. System analysis can often reveal component operations that are performed repetitively—and unnecessarily.

For component operations that remain essential, improvements in system performance can often be achieved by combining two, or sometimes three, operations, separately performed at as many work centers, into one combined task performed at a single center.

To make the point, imagine that operations P_2^1 and P_3^1 are combined in a single center, for convenience labeled P_{23}^1. If the processing time intervals $P_2^1 = 20$ minutes and $P_3^1 = 12$ minutes now require $P_{23}^1 = 33$ minutes in the combined arrangement, processing time and cost would be slightly higher, but $M_3^1 + {_3}D^1 + D_3^1 + M_4^1 = 8 + 209 + 20 + 3 = 240$ minutes would be saved, for an increase in system efficiency to

$$94/(25 + 477 + 95 + 80) = 94/677 = 0.139 = 13.9\%.$$

Whether, under these assumed conditions, the combination of P_2 and P_3 is desirable depends of course upon the increased cost of processing, compared to the values of faster service to customers and clients and lower costs of working capital, in-process inventory, and floor space. If $P_2^1 + P_3^1$ can be made equal to P_{23}^1, combining the two work centers into one *is* desirable—provided the combined operations are compatible.

As operational systems come into being and grow, a certain virtue is ascribed to division of labor, deriving perhaps from Adam Smith's famous description of pin making.[1] But in jobbing systems, dividing exacts penalties, and combining yields rewards. These can be stated axiomatically:

1. Adding a work center to a jobbing system by increasing division of labor requires moving intervals to and from the added center, and pre- and postoperation delays.

2. Eliminating an unnecessary operation saves not only the time hitherto required for processing but also the intervals that have been needed for moving work to and from the center and delays before and after processing.

3. Combining compatible operations for performance at a single center eliminates moving times and delay times previously required

1. Adam Smith, *The Wealth of Nations* (New York: Modern Library, 1937), pp. 4, 5.

for all but one of the combined centers. The net value of the combined operations depends upon the direct costs of processing, compared to the indirect benefits of improved service and reduced working-capital, in-process-inventory, and floor-space requirements.

Examples showing the consequences of eliminations and combinations of work centers are given in Chapter 4.

CONTROL OF JOBBING SYSTEMS

Control of any jobbing system, in the full sense of the word, requires current knowledge of

Capacity over future periods of time
Existing commitments
Priority of existing commitments
Prospective demands
Priority of prospective demands relative to existing
    commitments
Reports of completed and partially completed work

Raw materials also can be controlling, but their procurement will not be considered here.

For all the evolutionary progeny of jobbing systems, whole systems are the entities to be controlled, but for jobbing systems, control by work centers is necessary, since each center is, or is likely to be, different from others in the kinds of tasks to be performed and the time intervals required for performance.

This order is so large that it is seldom attained: the time and cost of controlling every work center would be more than control would be worth. It is therefore necessary to fall back upon selective control limited to key work centers.

This has been done in at least three ways: (1) by use of the historically important Gantt chart; (2) by use of the Gantt chart's more flexible successor, the machine-loading chart; and (3) by PERT and CPM techniques for the control of large one-time projects.

Gantt charts

Devised by Henry Laurance Gantt, the method was used with notable success in World War I to control the "turnaround" of merchant ships arriving in American ports to discharge inbound, and load outbound, cargo for England and France, then beleaguered by a German U-boat blockade. Elimination and reduction of delays (waiting

Fig. 3.1 A Gantt chart showing the state of the system at the time of review. The V at the top shows the review date, and the bold lines indicate the extent to which the planned program has been completed. All the jobs for which the bold lines extend beyond (to the right of) the V are ahead of schedule; those that fall short of the V are behind schedule by amounts of time proportional to unclosed distances. Reproduced from Robert H. Roy, *The Cultures of Management* (Baltimore: Johns Hopkins University Press, 1977), p. 127.

FOUNDRY — SIDE FLOORS

MOLDERS	FLOOR NO.	24 Mon.	25 Tue.	26 Wed.	27 Thurs.	28 Fri.	29 Sat.	31 Mon.	1 Tue.	2 Wed.	3 Thurs.	4 Fri.	5 Sat.
909 Conden	57	5144-4-2 6667W Pedestal											
			286-3-13042W Cylinder					341-6-13080 Cyl IPer Day					
		5-428-8-8640 W Piston - IPer Day											
		5228-12275 W - Pedestal				5462-7 8691 W-Piston- 2 per Day							
			300-3-1 3510 W										
901 Kelsey	58	7428-50-8045 OE Plate						170-50-8646 CE Plate					
866 Malone	58-2	238-8-13783 W		2 04-4-13 290 W Head				T26-25 12983 W Heads					
		282-5-13586W Piston		1231F 2-1328W				6744-20-12767 W-4 Pos Per Day					
		285-6-13879 W Head						287-24-13463 W-5 IPer Day					
			321-6-13315 W Head					13283 11424 Replace			6165-13285W		
965 Piper	59	3102-30-7644 W-Cover 2 Per Day											
		5486-40-8660W Head I Per Day											
		5295-8 940 W		5420-12-8978W Crosshead IPer Day							5430-4-4805		
								8540 Replace					
859 Richardson	59-2	271-70-3500 V Cover I Per Day											
		6269-12-8356 W Gear IPer Day								8648-14 261 W			
		7071-6-13315W Arm 2 Per Day			16131	8603-12-8350 W Gear 2 Per Day							
			1-14339W Threshold			3451-12-8836W					452-4-8363 W		
											481-4-8332 W		

for a berth at dockside, for stevedores and dock space for inbound cargo, for transport to carry it away, for outbound cargo, fuel, etc.) cut turnaround time in half, adding the equivalent of many ships to the Allied merchant fleet at a critical time.[2]

Figure 3.1 is a portion of a Gantt chart used to control the operation of the molding department of a foundry. The cells in the

2. L. P. Alford, *Henry Laurance Gantt* (New York: American Society of Mechanical Engineers, 1934), chaps. 14, 15.

12-day vertical columns and the horizontal rows for the five molders portray the capacity of the work center, diminished by estimation of delays and intervals during which operators are not available, blocked out by the crossed diagonals. Commitments forecast at the time of the original drawing are indicated by the beginning and ending brackets (⌈⌉).

At the date of review, shown by the V at the top, bold horizontal lines indicate the status of each job: closure to the right bracket (⌉) indicates completion, and partial closure shows the portion yet to be done. If the unfinished part lies to the left of the review date, that job is behind schedule, as shown for five of the seven jobs assigned to Molder Comden. Corrective action such as assigning additional manpower or overtime could, of course, follow.

Machine-loading charts

Once a forecast has been committed on a Gantt chart, a change in priority requires cumbersome redrawing, a serious disadvantage in jobbing systems where rearrangements frequently are imperative. This disadvantage may be overcome by using loading charts.

Loading charts take many forms but in general have the same rows and columns as Gantt charts. Their entries—on cards or pressure-sensitive tape cut to lengths proportional to time predictions— are easily placed and easily moved to new priority positions.

These techniques are in essence representations of inventories of work in process and in prospect. Thus, they are akin to well-established accounting methods for perpetual inventories of raw materials, spare parts, finished goods, and the like, and budgets forecasting permissible expenditures and encumbering restricted funds. It is not surprising, therefore, to find accounting procedures applied to jobbing system control. It is even less surprising to find computers adding new commitments to those already stored, obeying instructions to set or change priorities, diminishing or deleting obligations, and providing situation reports by means of print-outs or visual displays.

PERT and CPM

Control of very complex, for the most part one-at-a-time projects has been greatly improved by two similar computer-dependent methods: PERT, the acronym for Program Evaluation and Review Technique, was conceived by the Special Projects Office of the United States Navy in collaboration with consultants from Booz, Allen and

Hamilton, and, almost coincidentally, CPM, the acronym for Critical Path Method, was developed by engineers at the DuPont Company.[3]

Each method requires the design of networks like that suggested by Fig. 3.2, networks that are many times larger and much more complex than can be shown here. Following *S* for start, the circled numbers denote "events," defined as accomplishments at particular instants of time (e.g., completion of an operation). "Activities" portray the time or resources necessary to progress from one event to the next; in the enlarged section of Fig. 3.2 these activities are indicated by three numbers: an optimistic estimate of the shortest time in which the activity can be carried out, an interval judged most likely of attainment, and a longer, pessimistic forecast. The lengths of the activity lines are not scalar, as in Gantt and machine-loading charts. Rules forbid completion of an event until interdependent predecessor events have been completed and do not permit "looping" to lead back to a predecessor event.

As in the cases previously described, PERT and CPM require analysis: the conceiving and listing of every event, the mapping of all sequences and interdependencies, and estimates of all times and resource requirements. For complex, multistage projects this is a considerable undertaking, valuable in itself but even more valuable in application. Given these data, a computer can be programmed to search for, find, and identify the longest—that is, critical—path through the network. The sum of the most likely predictions, assuming that these are used, will then represent the interval between starting and finishing the project.

However, more than this can be done. If the critical path is the longest, then all the other paths must be shorter, and for each such path there will be "slack," intervals during which these paths will be idle, will have unused capacity. Redirection from the critical path to paths having slack can speed up the project and reduce working-capital requirements.

PERT and CPM methods have been applied extensively in the construction industry. Under the rubric "Construction Management," there are consultants who specialize not in building per se but in the technology of jobbing system control.

In each of these procedures the beginning of every event requires estimation of when that event is to be completed and provides feedback of progress that then guides corrective measures and ascertains

3. Harold Koontz and Cyril O'Donnell, *Principles of Management*, 3rd ed. (New York: McGraw-Hill, 1964), pp. 575ff.

Fig. 3.2 A PERT network. The lower figure is an enlargement of the lower left portion of the upper figure, with optimistic, most likely, and pessimistic time entries shown between each related pair of numbered events. (The numbers assigned to the events have no sequential meaning.) The critical path in the enlarged portion is indicated by the bold arrows. Reproduced from Roy, *The Cultures of Management*, p. 125.

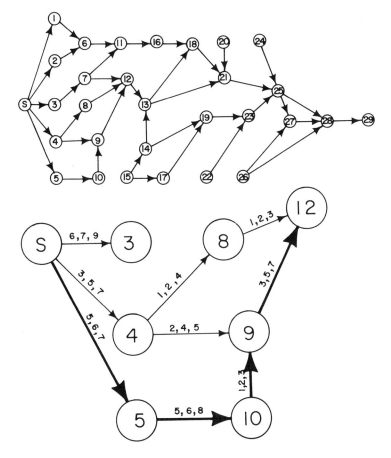

their effects. Of these universal precepts of control,[4] the one that is most important, the one upon which the others depend, is the first: that for an event beginning at time t there must be a forecast of completion at time $t + \Delta t$.

Gantt and machine-loading charts and PERT and CPM programs are too costly for many jobbing situations, but the $t + \Delta t$ concept

4. Gordon S. Brown and Donald P. Campbell, "Control Systems," in *Automatic Control* (New York: Simon & Schuster, 1955).

can be applied to key work centers, as the case histories of Chapter 5 will show.

PRODUCTIVITY IN JOBBING SYSTEMS

Compared to successor technologies, system efficiencies for jobbing are inevitably low because of delays preceding and succeeding operations and moves, all of which are included in the denominator of equation (1.4). System efficiencies, as has been shown, can be dramatically improved by eliminating unnecessary operations and reducing division of labor.

Operational efficiencies in jobbing systems (equation [1.3]) also are low, and the cause is often lack of control. Outputs, as measured in the numerators of equations (1.3) and (1.4), are reduced by human tendencies to relax and procrastinate in prevailing "climates of complacency," and inputs expressed in the denominators are at the same time increased, thereby reducing quotients in the output/input ratios.

These deleterious effects have serious economic consequences. Between 1970 and 1979 there was an increase of 1.4 million employees in manufacturing in the United States, a rise of 6.75 percent based upon 1970 figures. During the same period, the number of service and public administration employees combined rose by 7.6 million, a 30.69 percent increase.[5]

Using different but analogous categories of blue-collar and white-collar workers, Lester C. Thurow, Professor of Economics and Management at the Massachusetts Institute of Technology, reports that white-collar workers now outnumber blue-collar workers by 20 million. More pertinent is his finding that productivity in the American economy has been falling: the rise in manufacturing operational efficiency has been more than offset by the larger decline in the operational efficiency of the white-collar sector.[6]

Jobbing system technologies are more likely to be found in public administration, service, and white-collar occupations than in manufacturing systems, which are almost always manned by blue-collar workers. It is reasonable to conclude, therefore, that jobbing systems have increased in number, kind, size, and importance—to the detriment of national productivity.

---

5. U.S. Bureau of the Census, *Statistical Abstract of the United States, 1980* (Washington, D.C., 1980), table 679, p. 406.

6. *Newsweek*, August 24, 1981, p. 63.

Two means of improving the performance of jobbing systems that cannot be articulated, balanced, or otherwise transformed have been described: (1) elimination of unnecessary operations and diminution of division of labor, and (2) institution of control over key work centers.

Accomplishing these objectives would not be easy: measurements for forecasting are difficult, systems analysts are always suspect, the changes they propose are disrupting, habits resist change, job insecurity evokes fear, and complacency can become urgency only through compassionate, adroit, diplomatic, determined, and persistent effort.

Throughout the economy, particularly in the bureaucracies of every kind of formal organization, jobbing systems are as pervasive as they are costly. Improving them would be very difficult—but the game would be worth the candle.

FOUR

Jobbing System Simplification

PURCHASING
PUBLIC WORKS PROCUREMENT
A PEDIATRIC OUTPATIENT CLINIC

This chapter describes three jobbing sequences, each a subsystem of a much larger system. The performance of two of these was significantly improved by the elimination of unnecessary operations and by reduced division of labor. The third, a system related to public works procurement, was not improved by these means but could have been; it is described to illustrate the difficulties attending system simplification in a political-bureaucratic environment.

PURCHASING

Table 4.1 shows the routine followed for purchasing raw materials for an organization from which customers often demanded deliveries of finished goods within four weeks from the date of order.

Each new order, almost always different from preceding and succeeding orders, went first to a design and planning operation, where decisions were made about the raw materials required. Requisitions for these were written by hand and placed in an out-basket for delivery to a typist. Twice each day an interior mail clerk picked up these requisitions and placed them in the in-basket of a typist, who transcribed them and placed typed and handwritten copies in the out-basket for transfer back to the planner for his signature. The signed copy then went by the same means to the purchasing agent.

Authority to convert requisition to purchase order was vested solely in the purchasing agent; he could approve the planning specification or, if he believed that he could buy more advantageously, change the material to be ordered. In either case, the purchase order was initialed by the agent, again in handwritten form. Delay and processing intervals were longest at this stage.

30

Table 4.1 Flow process chart of the original routine for purchasing raw materials.

M_1	Move job to planning.
$_1D$	Await planning.
P_1	Plan production; prepare handwritten requisition for materials; place requisition in out-basket.
D_1	Await move to typist.
M_2	Move to typist and place handwritten requisition in in-basket.
$_2D$	Await transcription.
P_2	Type requisition and place it in out-basket.
D_2	Await move to planning.
M_3	Move to planning.
$_3D$	Await signature.
P_3	Sign typed requisition and place it in out-basket.
D_3	Await move to purchasing.
M_4	Move to purchasing; place signed requisition in in-basket.
$_4D$	Await preparation of purchase order.
P_4	Prepare handwritten purchase order and place it in out-basket.
D_4	Await move to typist.
M_5	Move to typist and place handwritten purchase order in in-basket.
D_5	Await transcription.
P_5	Type purchase order and place it in out-basket.
D_5	Await move to purchasing.
M_6	Move to purchasing and place typed purchase order in in-basket.
$_6D$	Await signature.
P_6	Sign purchase order and place it in out-basket.
D_6	Await distribution.
M_7	Distribute copies of the purchase order.

NOTE: Because all delays ($_iD$ and D_j) are assumed to be positive, there is no idle time ($I_j = 0$) at any of the six work centers.

Again, the handwritten form was conveyed by interior mail to a typist, transcribed on to a multicopy form, and returned via the same route to the purchasing agent. His signature gave necessary authority for the purchase by transmittal to the vendor, if need be by telephoned order, confirmed by mail.

Observations and interviews revealed that these incredibly cumbersome procedures often consumed half of the allowed schedule intervals, creating chaos and costly overtime in the factory, strained relations with customers, and distress in the sales department.

Desirable changes, shown in Table 4.2, are almost too obvious to recount. All typing and delays deriving from these operations

Table 4.2 Flow process chart for revised routine for purchasing raw materials.

M_1	Move job to planning.
$_1D$	Await planning.
P_1	Plan production, prepare and sign handwritten requisition for raw materials, and place requisition in out-basket.
D_1	Await move to purchasing.
M_2	Move to purchasing and place requisition in in-basket.
$_2D$	Await signature of purchasing agent.
P_2	Sign purchase order, mark for distrubution, and place it in out-basket.
D_2	Await distribution.
M_3	Distribute copies of the purchase order.

NOTE: Operations P_2 and P_5 from Fig. 4.1 have been eliminated, and Operations P_1 and P_3, as well as P_4 and P_6, have been combined. Very large gains in cycle times and system efficiencies resulted from the elimination of eight intervals of delay.

were eliminated as unnecessary, in favor of handwritten requisitions on multicopy purchase order forms. These were prepared by planners and sent by interior mail or carried directly to the desk of the purchasing agent. His signature on the same form converted requisition to purchase order, a copy of which was then transmitted or telephoned to the vendor. Should the purchasing agent wish to change the specification, he could do so by preparing a new handwritten form.

These changes were so obviously desirable as to bring them about without measurement of the time intervals for the moves, delays, and processes of Tables 4.1 and 4.2. Results of the change were salutary: reduction in time required for procurement by more than half, relief from crises that had become chronic, and wage and salary economies from needless typing and from spot-overtime in the factory.

Opposition to the change came from the purchasing agent, who felt threatened with loss of authority by the shortened procedure. He could not block the proposed change, but he was able to negate one additional proposal: that planners be authorized to buy raw materials costing less than a specified amount, subject to designation of approved vendors by the purchasing agent. Opposition of this kind to system changes is common and often difficult to overcome.

Materials may have been purchased advantageously by means of the original method in terms of price and quality. In that sense, purchasing may have been regarded as effective. But even if this were so, whatever effectiveness may have been accorded to the tedious and time-consuming procedure, it was suboptimal, achieved at a net loss in effectiveness of the system as a whole.

PUBLIC WORKS PROCUREMENT

A much more complex jobbing sequence, one that may well typify a great many bureaucratic operations, was found during the deliberations of a Construction Investigation Committee appointed by the mayor of the city of Baltimore.[1]

The committee was appointed to investigate alleged construction defects, either unreported or uncorrected in the construction of a school. Charges were made public by members of the Building and Construction Trades Council, AFL–CIO, against the nonunion contractor, and the committee was appointed at the instigation of the city comptroller.

At the initial hearing four witnesses from user departments[2] testified. All of them preferred to discuss not construction defects per se but the cumbersome, protracted, multistage, system by which public works are carried from conception to completion. Similar testimony was given at a later hearing by representatives of the Associated Builders and Contractors, Inc. There were repeated declarations that numerous reputable architects, engineers, and contractors avoided city work because of system complexities.

Sensing possibilities for improvement, the committee asked for and received permission to examine procurement procedures, and enjoyed the services of a staff analyst made available by the director of the Department of Finance.

Flow charts were made to show sequences and alternatives for sequences other than construction itself.[3] These were too large and detailed for illustration here, but several sequences can be described

1. This account is based upon the author's experience in 1979–80 as chairman of a Construction Investigation Committee for the city of Baltimore, previously reported in different context in "Politics, Social Forces, and Engineering Management," *Engineering Management International* 1, no. 1 (1981): 13–16.

I am very much indebted to the late Joseph M. Axelrod, a member of the Construction Investigation Committee and former deputy director of public works for the city of Baltimore, for details set forth here.

2. These were spokesmen for the Departments of Recreation and Parks, Education, Finance, and Housing and Community Development.

3. The operations of construction itself will not be exemplified here, but the Construction Investigation Committee did recommend that the Department of Public Works apply techniques of "construction management" of the kind described for

to portray the excessive division of labor that prevailed and the inevitable delays that occurred before and after each stage.

Typical procedure following budget approval by the Board of Estimates[4] is a request from the user agency to the Consultants Evaluation Board for a list of approved professional firms to be considered as architects and engineers for the project. At approximately the same time, the Architectural and Engineering Awards Commission provides the user agency with a list of firms that have not been awarded work on the last four projects "in that category." Based upon that list and the approved list received from the Consultants Evaluation Board, the user agency prepares a "short list" of from three to five firms, arranged in order of preference. The short list is returned to the Architectural and Engineering Awards Commission, approved or altered by it, and returned to the user agency with a recommendation as to the mode of negotiation (competitive design, competitive bidding, or negotiated fee).

The user agency then arranges for presentations from each firm on the selected list and, except when competitive bids are recommended, makes a selection and negotiates a contract. When competitive bids are called for, the Bureau of Purchases secures these and they are opened by the Board of Estimates. The contracts that emerge from these procedures are screened by the Department of Audits and, when found satisfactory, again go to the Board of Estimates for approval. Only then can services for design and engineering be retained.

Intervals of time for these steps vary widely, with a six-month lapse roughly typical *following* budget approval. Most of these stages are sequential, each characterized by delays of the kind ascribed to every kind of jobbing system.

Given that sound judgments can be made about probable budget approval of many public works projects, most of these predesign stages can be carried out concurrently rather than sequentially, during the interval between budget proposal and approval. This was recommended by the Construction Investigation Committee as a

PERT and CPM programs (see Chapter 5). This recommendation was also made by the Mayor's Committee.

4. Under the chairmanship of the president of the City Council, the Board of Estimates is composed of the mayor, the comptroller, the city solicitor, and the director of public works. The first three of these are elected, the other two are appointed. The board is the most important executive body in city government; it meets for all or part of one day each week to consider many pages of agenda. Many items are of little consequence and are given only perfunctory attention, but the procedures—and the "before" and "after" delays—continue.

means of shortening the time requirements by several months during the beginning stage.

Combined operations

In addition to the Consultants Evaluation Board and the Architects and Engineering Awards Commission, there was also a Contractors Prequalification Committee. These three bodies performed different but related tasks, all preliminary to the engagement of architects, engineers, and contractors. At the design stage there were two analogous groups: a Design Advisory Panel and a Commission for Historical and Architectural Preservation. All five groups met at intervals, each contributing a useful function but also contributing to system delays. The Construction Investigation Committee did not recommend elimination of function, but did propose combining the first three and the last two into two single commissions.

Change orders

Public works projects, even small ones, often require unexpected changes from original designs and specifications, and sometimes such needs and the costs incidental to them must be decided forthwith, lest further work be brought to a halt.

To cope with these contingencies, when the project supervisor (a city employee) and the contractor agree that a change is necessary, the contractor estimates the extra cost, a "change order" is prepared and signed, then signed by the project supervisor, and finally sent to the division head for his scrutiny and his signature. The change order is then sent to the appropriate bureau in the user agency, again for scrutiny and signature. Thence it goes to the Department of Public Works, technically for final authorization by the director himself, but more often for the signature of his deputy.

There are many such change orders, for each of which the deputy's signature makes him responsible, even though he cannot possibly be familiar with many of the changes he is asked to approve. By then, much of the work has already been completed. Thus, he is confronted each day with a stack of dilemmas: he can sign without careful scrutiny and incur risk of retribution, or he can investigate to whatever extent is possible—and have time for little else.

The course of a change order through these steps takes from one to three weeks, but the end is not yet in sight: payment for the change order requires progression through the Office of the Comptroller, audit if the amount is larger than a specified limit, entry upon the agenda of the Board of Estimates, approval at a weekly meeting of that body, and thereafter submission to the disbursement division of the city treasurer's office.

Auditors are properly motivated to save money for the organizations they serve, so it is not altogether surprising that some change orders that have been approved in terms of scope and amount, and for which the work has long since been done, are somewhat forcibly renegotiated, confronting the contractor with an invidious choice: accept less or take legal action.

Disbursements, even after approval by everyone in the prescribed operational sequence, are sometimes made within the customary 30-day credit interval, but many accounts remain unpaid for much longer times. Egregious delays pervade not only this small segment of public works procurement but the entire process, giving rise to credible testimony that some architects, engineers, and contractors eschew city work because of these onerous and protracted complications.

Remedial proposals for thorough analysis of the procurement system—proposals to reduce the number of signatures required on change orders, to relieve the Board of Estimates of the burden of *pro forma* approval of trivial items, to use random sample auditing in ascending frequencies gauged to dollar values, and to adhere to accepted commercial practices in paying bills—were well received by the Board of Estimates, and the Construction Investigation Committee was asked by the mayor to remain active "to implement the changes recommended."

The Mayor's Committee on Building Design and Construction Procedures

Concurrently with the deliberations of the Construction Investigation Committee, a second committee having analogous purpose also was sitting. The Mayor's Committee on Building Design and Construction Procedures consisted of five city officials and three corporation executives:

> Director of the Department of Public Works, chairman
> President of the City Council
> City Comptroller
> City Solicitor
> Director of the Department of Finance
> President of a large construction company
> President of a consulting organization
> President of a remodeling company

The first four city representatives were members of the Board of Estimates, and all three of the corporate members presided over organizations frequently engaged in public works projects for the city.

The Construction Investigation Committee was requested to maintain liaison with the Mayor's Committee and attempted to do so, but without success. None of the proposals for system analysis, elimination of unnecessary operations, reduction in division of labor, or use of sampling techniques was adopted, nor were any of these recommendations included in the 14 items listed in the report of the Mayor's Committee. One of these items did urge use of construction management techniques, but many of the other proposals could only be regarded as expressions of wishful thinking.[5]

The rationale for continuing system complexity and for lack of responsiveness to proposed system analysis requires conjecture.

Political systems and the bureaucracies that are part of them seek the appearance of integrity. Multistage sequential and time-consuming consideration of who shall design, engineer, and construct a public work *appears* to ensure legitimacy. Multiple signatures on change orders *appear* to verify that each is correct, an appearance that is then reinforced by auditing that *appears* to be meticulous. It is probable, perhaps even provable, that the kinds of concurrence, consolidation, elimination of unnecessary operations, and random sampling procedures proposed by the Construction Investigation Committee would enhance system integrity, though perhaps at some perceived lessening of safeguards.

Speculation that the appearance of rectitude has very high value is reinforced by the recent history of political institutions in the state of Maryland and the city of Baltimore. A former county executive and governor was forced to resign as vice-president of the United States; another governor was sent to prison; two other county executives also were sent to prison; so was a president of the City Council (and member of the Mayor's Committee). In the Department of Public Works, a former deputy director was convicted for various offenses, along with demolition contractors; another department employee pleaded guilty to charges of selling a city-owned

5. "Report of the Mayor's Committee on Building Design and Construction Procedures," January 9, 1981. Among the recommendations characterized as wishful were these: "4. Construction specifications should be complete and concise.... 6. Design consultants and construction contractors should be held rigidly to the completion dates specified in their contracts.... 8. Monthly progress and final payments to contractors should be prompt and in accordance with the terms of their contracts.... 10. Change orders must be processed within a reasonable time."

front-end loader to a "sting" operation set up by the police. On a lesser scale were indictments of six city building inspectors for bribery.[6]

As a consequence of frustration over what was felt to be a cold shoulder from the Mayor's Committee, the Construction Investigation Committee came to believe that system analysis was feared—certainly not wanted. It was also the latter committee's perception that the Mayor's Committee was not suitably constituted to evaluate, in effect, its own performance, there being a considerable degree of self-interest on the part of most members of the group.

From this example, and from these conjectures, one conclusion can be drawn: analysis of the jobbing systems that pervade government at all levels has the potential to yield very large gains to the economy, but inertial forces—lack of motivation, self-protection, preservation of the status quo, and a certain amount of fear—also are very powerful.

The public works procurement system, like so many in the bureaucracies of government, was—and perhaps still is—egregiously inefficient and very ineffective.

A PEDIATRIC OUTPATIENT CLINIC

An outpatient clinic, dedicated to the diagnosis and treatment of children up to the age of fourteen, provided health care to large numbers of patients, many of modest means, some best described as indigent.[7] Some patients first came to the clinic or returned there by appointment; others came unannounced. Most of the children were accompanied by mothers, members of the family, or friends,

---

6. All of these cases were reported in the press. Details can be found in the *Baltimore Sun* on or about the following dates: October 11, 1973 (resignation of Vice-President Spiro Agnew); August 24, 1977 (conviction of Governor Marvin Mandel); March 3, 1974 (conviction of County Executive Dale Anderson); January 4, 1975 (sentencing of County Executive Joseph Alton); September 22, 1982 (guilty plea by President of the City Council Walter Orlinsky); October 29, 1977 (conviction of Deputy Director of Public Works Ottavio Grande); July 24, 1980 (conviction of City District Superintendent of Highway Maintenance Charles Delivuk, Jr.); September 11, 1981 (indictment of six city building inspectors). See also Bradford Jacobs, *Thimbleriggers: The Law v. Governor Marvin Mandel* (Baltimore: Johns Hopkins University Press, 1984).

7. I am indebted to a former graduate student, Dr. Maurice M. Taylor, for this example, excerpted from his Master's essay, "Analysis of Congestion in an Outpatient Clinic" (Johns Hopkins University, 1958).

so that the number of people in the system at any given time—the in-process inventory—was almost double the number of patients.

The illnesses of childhood, ranging from accidents to tumors, were treated in an operational sequence that involved a registrar, an admitting officer, a screening doctor, and a cashier. There were also nurses to weigh patients and take temperatures, and physicians to diagnose, treat, and prescribe. Frequently, case histories would be called for from a history room located outside the clinic; when found by the history clerk, these were delivered to the clinic by a messenger. Prescriptions required patient visits to the pharmacy, located elsewhere in the hospital.

The registrar's station was the almost invariable point of entry to the system. There patients were provided with necessary papers to fill out, and there case histories for repeaters were called for. The admitting officer had the somewhat delicate task of ascertaining ability to pay and assessing appropriate fees. The screening doctor, usually an intern, weeded out the needs that were less pressing from those of greater urgency and routed patients to medical areas judged suitable by the screening diagnosis. The cashier had the obvious duty to collect the fees according to information supplied by the admitting officer.

Routes taken through the system by 117 patients on a test day are shown in Table 4.3. There were 46 different operational sequences, replete with backtracking that in some cases reflected patient confusion as to where to go next.

Between each pair of successive operations in the 46 rows of Table 4.1 were waits and moves to which we have devoted so much attention. In this example the in-process inventory was people, who moved to the next operation immediately when finished at a given work center; waiting time before moving was therefore of little consequence, nor, except for confusion as to where to go, was moving time or distance. But at each successive station, each patient had to join the queue already there; his or her waiting time depended upon the length of the queue and the service time of the work center. On occasion, when a patient did reach first position in a medical area, the history would not be there, and he or she would be sent back to wait again.

The contrast between total time spent in the system by patients and the services provided to them on a test day is illustrated by the histograms of Fig. 4.1. These data show an average service time per patient of 32.5 minutes and a mean total waiting time of 87.3 minutes. From the data in Table 4.3, the average number of work centers visited by each patient was calculated to be 4; these numbers thus

Table 4.3 Routes taken by patients on a test day

No. of patients	Work center route	No. of patients	Work center route
2	C	4	R D C M
1	D	1	R D C M C
1	D M D M R	1	R D C M P C
3	R	2	R D C M R
1	R A	1	R D C M R A C P
1	R A C	1	R D C M R C M
16	R A C M	1	R D C M R C P
1	R A C Secretary	2	R D C R A C P
1	R A D R C M R A C	2	R D M
1	R A R P	1	R D M R A C P
3	R M	1	R D M R D R M
3	R C M	2	R D R
1	R C M P	2	R D R A C
1	R C M R	13	R D R A C M
6	R D	3	R D R A C M C
7	R D A C M	1	R D R A C M C P
2	R D A C M C	11	R D R A C M R
1	R D A C M C P	1	R D R A C M R C
5	R D A C M R	2	R D R A C P
1	R D A C M R A C	1	R D R A M
1	R D A C M C R A C P	1	R D R A P
1	R D A C M R C	1	R D R C M R
1	R D A D R	1	R D R C P

NOTE: Steps required to secure and deliver case histories are not shown. Symbols representing work centers are: R = registrar, A = admissions officer, D = screening doctor, C = cashier, P = pharmacy, M = medical area (the services of nurse and attending physician are combined).

suggest service of about 8 minutes at each station and a waiting time of almost 30 minutes at each paired interface.

Detailed information about system performance over time is always tedious, disruptive, and difficult to obtain, but the above numbers, though meager and subject to large variances, were sufficient to provide objectives for the study at hand: to make the system less time-consuming and confusing to patients and their often apprehensive escorts, to relieve congestion that was harassing to the staff, and to do so without increasing cost or diminishing the variety or quality of medical care.

The objectives were for the most part realized by means of

Fig. 4.1   Distribution of waiting and service times for a test day. Adapted from Maurice M. Taylor, "Analysis of Congestion in an Outpatient Clinic" (Master's thesis, Johns Hopkins University, 1958).

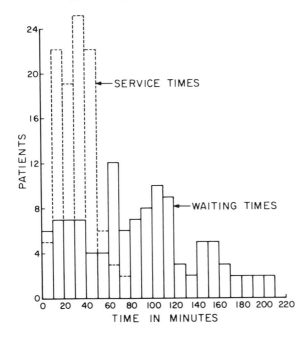

changes related to division of labor. The concept of parallel queues was used by assigning a second registrar and giving to both the duties previously performed by the cashier, whose separate function was thereby eliminated. The job of the messenger who brought histories to the clinic also was eliminated, by installation of a dumb waiter connecting the history room and clinic. Trips by patients to the distant and hard-to-find pharmacy were reduced by keeping more frequently prescribed medications in the clinic itself.

The effects of these changes are shown in Table 4.4, a summary of a Monte Carlo simulation of the passage of 40 patients through the clinic.[8] The arrival time of each patient and his or her ordinal number are the same for both existing and planned procedures.

8. As in the game of roulette, from which the term derives, *Monte Carlo simulation* involves probability distributions for occurrences of events, in this case arrivals, work center sequences, treatments, departures, etc. For a simple, easily played example using two ordinary dice, see Charles D. Flagle, "Queueing Theory," in *Operations Research and Systems Engineering,* ed. Charles D. Flagle, William H. Huggins, and Robert H. Roy (Baltimore: Johns Hopkins Press, 1960), pp. 403ff.

Table 4.4 Monte Carlo simulation of the passage of 40 patients through the outpatient clinic

Time (in min.)	Existing procedure					Planned procedure			
	Registrar	Doctor	Admissions officer	Cashier	No. who finished	Registrar & cashier	Doctor	Admissions officer	No. who finished
5	0	0	0	0	1	0	0	1	1
10	0	0	0	0	4	0	0	0	4
15	3	0	0	0	6	2	0	0	6
20	6	0	0	0	6	3	0	1	7
25	8	2	0	0	9	3	2	2	10
30	10	0	0	0	10	0	1	4	13
35	8	0	1	0	10	0	0	0	15
40	7	0	1	1	11	0	0	4	19
45	4	0	4	4	13	0	0	5	20
50	7	0	4	1	19	0	1	5	23
55	6	0	3	0	21	0	2	4	26
60	6	0	5	0	23	1	3	3	28
65	3	0	5	1	25	0	4	1	32
70	2	3	4	0	27	0	1	1	33
75	0	3	2	1	30	0	0	2	37
80	0	0	5	0	34	0	0	0	40
85	0	0	2	0	36		Finished		
90	0	0	0	0	40				
			Finished						
Total number in queue					122				62

SOURCE: Adapted from Maurice M. Taylor, "Analysis of Congestion in an Outpatient Clinic," pp. 90–94.

Improved system performance is reflected by a 10-minute gain in the passage of the assumed patient batch through the clinic (an improvement of 11 percent), and by a 50 percent reduction of the aggregate number waiting at work centers. Conditions for patients have indeed been made better, as have conditions for clinic personnel, now less harassed by the visible presence of sick children and anxious parents.

The quality of medical care provided to the patients of the clinic was acknowledged to be good. Under the more efficient arrangement the effectiveness of medical care was in no way impaired. Given the less crowded conditions and the quicker passage of patients through the system, effectiveness as well as efficiency probably increased.

FIVE

Jobbing System Control

THE COMPOSITION OF TYPE
LEGAL AND JUDICIAL PROCESSES
NUCLEAR POWER

Because of variations of the kind described in Chapter 2 between different jobs and different work centers, control of jobbing systems has been said to require surveillance at *every* work center. For most jobbing systems this stipulation is too demanding to be practicable. Instead, control must be exercised only at "key" work centers, with due regard for system capacity, existing commitments, prospective demands, priorities, and accomplishments, the components of control specified in Chapter 3.

THE COMPOSITION OF TYPE

The following example describes a subsystem that is obedient to these precepts.

In a printing plant that specialized in the manufacture of scientific books and periodicals, there were five work centers, each consisting of multiple work stations devoted to the composition of type. These were:

Copy preparation (redaction)
Keyboarding
Casting
Proofreading
Hand composition

There were also ancillary operations, but these need not be considered here.

Such control as existed was exercised by use of "mailers" that gave addresses to which proofs and copy should be mailed and dates by which these should be dispatched. System demand could be approximated by the size of the stack of mailers, but there was no satisfactory way by which to measure system capacity. Mailers were continually shuffled into new orders of priority and, despite much expediting, late mailings were the rule. To expedite work that lagged too far behind, or about which customers had complained, "RUSH" stickers were affixed to high-priority jobs, but there came to be so many of these "wolf cries" as to destroy their intended purpose. Chronic lateness, as in so many jobbing systems, became a way of organization life, uncomfortable to some but acceptable to most.

The method of control by use of mailers was deficient in another important way: the mailers remained at the work center from which proofs were mailed, where they were accessible to the scheduling supervisor, but they did not specify the order of work to be followed at any of the many work stations in the system, at each of which there was an in-process inventory of the kind that is characteristic of most jobbing systems.

In an attempt to come to grips with these difficulties, plant analysts first tried to measure system capacity, not only in terms of available employee hours but also in terms of physical units of output. This problem was confounded by extreme variances between jobs. A large book, for example, might be estimated to require 600 keyboard hours, with correspondingly heavy demands upon each succeeding work center. At a lower extreme a letterhead might require only a few lines of type, use no keyboarding or casting, and require but a few minutes of hand composition and proofreading. Each, however, would be considered a separate job, and to each a job symbol and number would be assigned for identification and accounting purposes.

To reduce such extreme variances to a more tolerable level, it was decided to divide keyboarding into "takes," each estimated to require 20 keyboard hours or less, and each counted as one item in the approximation of capacity. Between and among items there were still large variances, but given a reasonably stable product mix, expression of system capacity in items per day proved to be practicable. For purposes of discussion, capacity will be assumed to be 50 items per day, a number that could, of course, be varied to accommodate holidays, vacations, new hires, and the like.

During copy preparation each job was divided into takes in the manner described and each take was separated into its component parts (text, inserts, footnotes, legends, tables, display). To each of these components letters were assigned in alphabetical order, *a, b, c, d,* and so on, with *x* indicating the final component. Copy for each

Fig. 5.1 Cover of a take envelope coded 8/23-10 *cx*. The job symbol "BC Sept." identifies the September issue of the *Journal of Biological Chemistry*. Reproduced from Robert H. Roy, *The Management of Printing Production* (Washington, D.C.: Printing Industry of America, 1953), p. 90.

SFCOMPENV—FRONT

TAKE NO. 8/23 - 10 cx **CK CJ CA CM NO.** 318

Job Symbol BC Sept. **Job No.** 564

TAKE:

Class of Matter:

- ☐ Text
- ☐ Inserts
- ☐ Display
- ☑ Tabular
- ☐ Formulas
- ☐ Running Heads
- ☐ Ads
- ☐

TAKE LETTER	TYPE SIZES AND FACES: FACE		MOLD	NO.	WITH	CK MEASURE	ARRANGE-MENT	CK CLASS-IFICATION	TIMES TO CAST	EST. HOURS
a	10	on	12	8		32-7	C	I	1	5
b	8	on	10	8		40-9	C	III	1	2
cx	6	on	6	8		54-0	C	VI	1	3
		on								

Collate { By copy after reading ☐ / Before reading ☑ } COG Classification 2

Mail Desk: Number galleys before collation { Yes / No ✓ }

Special Instructions

Prepared by EL

DISPLAY:

Description: ______ Page. Make up to ______ x ______ picas.

☐ Make changes indicated.

☐ Put on running head and footline.

Type Purchase { Ordered / Order enclosed ☐ }

ALTERATIONS { Galleys ☐ / Pages ☐ } Machine Comp. { Yes ☐ / No ☐ } Classification

MAKEUP:

ARTICLE No.	GALLEY Nos.	No. OF CUTS	No. OF TABLES	No. of PAGES	ARTICLE No.	GALLEY Nos.	No. OF CUTS	No. OF TABLES	No. OF PAGES

Pages ____ to ____ Done by ____

Pages ____ to ____ Done by ____

Pages ____ to ____ Done by ____

Pages ____ to ____ Done by ____

PROOF PRESS INSTRUCTIONS:

GALLEY PROOF: 7 x 24	11 x 24	PAGE PROOF: 7 x 9	8½ x 11	AD PROOF: COATED AD STOCK	REGULAR	TOTAL
3						

→ IF THIS IS NOT SCRATCHED OUT PULL ONLY ONE SET OF PROOF

MAILING INSTRUCTIONS:

MAIL TO	COPY	ORIGINAL	DUPLICATE
OUT	✓	1	2
G20			

RECORD WORK DONE ON THE REVERSE SIDE

No. 318

Fig. 5.2 Portion of a daily Composition Schedule Sheet. The form for each day contained 50 priority numbers, the capacity assumed for this description. Reproduced from Roy, *The Management of Printing Production*, p. 92.

COMPOSITION SCHEDULE SHEET

FOR___8/23_____

No.	Date En-tered	Date Com-pleted	Job Symbol	Takes — Take No.	Takes — CK Hours	CA No.	CJ No.	CM No.	Remarks
~~1~~	8/18	8/22	~~JU Aug~~				560		~~Ads~~
~~2~~	8/17	8/21	~~ASTM~~			501			~~Pages~~
3	8/22		Soel-Aug				610		Cover
4	8/10		Duet-Sept	a-dↄ	15				
~~5~~	8/10	8/21	~~Blood~~			1014			~~Cover~~
6	8/12		JCP July	ax	5				Inserts
7	7/30		XX nav	a-ex	20				
8	8/3		HW&D				710		Ads
9	8/3		Wiley					711	
10	7/29		BC Sept	a-cↄ	10				
11	8/6		L.B. Co			780			Gal.
12	8/9		Drug Bal.			803			
13	8/12		Best 8 T	a-dↄ	16				
14	7/28		Transit					693	Report
15	8/3		Ray Rept			715			Pages
16									
17									
18									
19									
20									

Fig. 5.3 Composition Completion Report indicating work released for mailing or for presswork. The headings "Red," "Green," and "White" are explained in the text. Reproduced from Roy, *The Management of Printing Production*, p. 92.

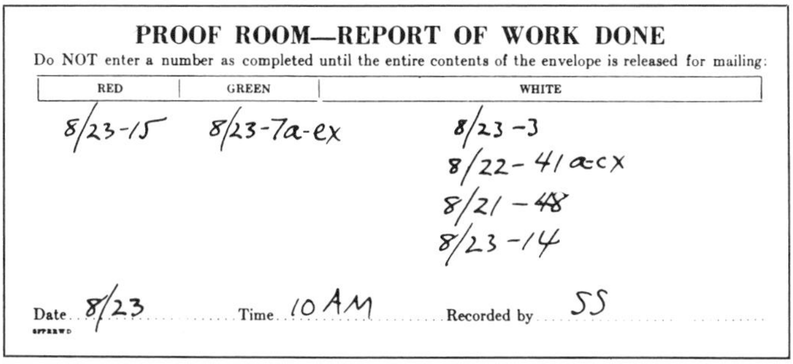
PROOF ROOM—REPORT OF WORK DONE

Do NOT enter a number as completed until the entire contents of the envelope is released for mailing:

RED	GREEN	WHITE
8/23-15	8/23-7a-cx	8/23-3
		8/22-41 a-cx
		8/21-48
		8/23-14

Date 8/23 Time 10 AM Recorded by SS

component would be put into a separate envelope, one of which is shown in Fig. 5.1.

All takes would then go from copy preparation to the scheduling supervisor, who, depending upon his knowledge of the state of the system and of relative priorities, would assign to each take a unique number, specifying not only the date by which proof should be mailed but also the order in which work should be done at each work station. Thus, Take No. 8/23-10 *a c x* told every supervisor and operator that this take was to yield precedence to all that had an earlier date or number, and take precedence over all that carried a later date or number.

The date and number of this take was then entered by the scheduling supervisor on the Composition Schedule Sheet for August 23, as shown on line 10 of Fig. 5.2. Sheets like this one contained 50 spaces for priority entries for each day, and there were sheets for as many days ahead as needed and also for days that had passed without completion of every scheduled item.

Feedback to the scheduling supervisor was provided by Composition Completion Reports like the one shown in Fig. 5.3. A line would then be drawn through each completed item, as shown on line 5 of Fig. 5.2. Supplemental feedback was also provided by a daily Composition Behind Schedule Sheet (Fig. 5.4) showing delinquent work, about which more will be said later.

The above method provided for all the control requisites specified in Chapter 3: an estimation of system capacity over future periods of time, existing commitments in priority order, means for

Fig. 5.4   A daily Composition Behind Schedule Sheet prepared by the composing room foreman. Reproduced from Roy, *The Management of Printing Production*, p. 93.

Production Control—Scheduling

COMPOSITION BEHIND SCHEDULE LIST   Date___8/24___

Take No.	Symbol	Take Letters	Caster	CMB, CU CJ	Reading	COG	Ready for CL	CL	Revision	Finals & Final Rev.
8/19-31	Chem Rev	a-dx		dx		a+b	c			
8/19-24	George									✓
8/20-42	State Rd.								✓	
8/23-21	BJH	ab x						✓		
8/23-37	R+H	ax			ax					
8/23-42	May	a-dx			a		b-dx			

fitting new demands into the scheduled array,[1] and reports of completed, partially completed, and delinquent work.

Despite its evident advantages over the previous method of control and its conformance with the above-stated precepts of control,

1. Knowledge of the frequencies of publication of periodicals and occasional advance knowledge of manuscripts for monographs permitted the encumbrance of some demands.

the new technique proved difficult to implement. There were two principal problems: one psychological, the other operational. The first of these was much more difficult than the second.

The earlier method of using dates on mailers to prescribe the order in which work should be done took little account of whether jobs could be done by appointed times, nor were the jobs that were to be mailed on any given day differentiated. Many of the dates were expressions of wishful thinking, and the use of "RUSH" stickers was so extensive as to create ennui rather than haste.

Old habits are hard to break. Accustomed as everyone was to a plethora of jobs behind schedule, the scheduling supervisor for a time ignored the 50-take capacity limitation by drawing additional priority lines on the back of each day's Composition Schedule Sheet, there assigning priority numbers as high as 75. The Behind Schedule Sheet soon came to list 300 takes, about half the total number in process.

Discussion of the problem with the scheduling supervisor revealed two beliefs: (1) that the schedule should call for what the system *ought* to do rather than what it *could* do; and (2) that the system would perform better—that is, at a higher rate of productivity—if it were continually goaded to "catch up." Behind these reasons, if one may call them that, lay an unthought-of tacit assumption: that in terms of customer relations, it is better to promise early and deliver late than to state and abide by what *can* be done. Prudent managers know that the second is the better choice.

Resolution of these problems required protracted training, during which all participants learned that there would always be late arrivals that properly demanded quick service—take numbers with early dates and preferential priority numbers. All also learned by experience that reservations—slack time—had to be made for these contingencies.

Additional flexibility was provided by color coding (see Fig. 5.3). A take number entered in red carried with it the rule that at every work station a red number must take precedence over all other numbers except another red take of higher priority. Use of green take numbers indicated that extra importance was attached to mailing those takes on the designated day, and gave precedence to green takes over higher-priority regular takes on that day. It was necessary for system analysts to insist that these preferential codes be used with restraint.

Occasionally, sometimes at customer behest, a specified completion date would have to be postponed; this could be indicated by simply striking through the take on the Composition Schedule Sheet for the original date and renumbering the take on the Schedule Sheet for a later date and on each component envelope.

By dint of these measures, the Behind Schedule Sheet was reduced to but a few delinquent takes. Lateness had been converted from rule to exception. The greatest gain, perhaps, was credibility, within the organization and among its customers.

Operational efficiency was not affected significantly by these controls, but system efficiency and system effectiveness were markedly improved.

LEGAL AND JUDICIAL PROCESSES

Legal and judicial processes[2] are not amenable to the kinds of prediction and performance measures just described, nor can procedures be divided into takes to reduce variations in work requirements. Nevertheless, control, and with it the administration of justice, has been improved by a remarkable federal law that provides and enforces $t + \Delta t$ limits of the kind described earlier.

In 1976 the United States Congress passed, and the President signed into law, the Speedy Trial Act, landmark legislation to which the late Senator Sam Ervin of North Carolina devoted a rare combination of legal, political, and creative talent.[3]

Like most laws, this one is replete with "ifs" made necessary to encompass the immense variety of possible circumstances attending criminal procedures. It would be possible to trace all of these in process charts and decision trees, but doing so would exceed both my capability and purpose. I must therefore simplify by tracing a less complex path: the events attending arrest and conviction for a felonious federal offense.

From among many possibilities, the following sequence of operational events has been chosen:

1. Complaint
2. Arrest warrant
3. Execution of the warrant
4. Initial appearance

---

2. I am very much indebted to Frederic N. Smalkin, Esq., Magistrate for the United States District Court for Maryland, who provided me with the system events described here and with a copy of the Speedy Trial Act, cited in footnote 3.

3. Speedy Trial Act, Pub. L. No. 93-619, *as amended* 18 U.S.C. §§3161 et seq. (1976).

5. Preliminary examination
6. Indictment
7. Arraignment
8. Pretrial conference
9. Trial
10. Presentence investigation
11. Sentence
12. Appeal, incarceration

All these steps are, to a degree, well understood, but each will be explained briefly, with particular attention to items 5, 6, and 7, where emphasis has been added to the $t + \Delta t$ limits prescribed by the act.

1. *Complaint.* "The complaint is a written statement of the essential facts constituting the offense charged."[4] In the present discussion it is assumed that the alleged offender has been identified and that the complaint has been made by an officer and sworn to by him before a magistrate.

2. *Arrest warrant.* Upon finding probable cause that an offense has been committed by the defendant, the magistrate issues a warrant which commands that the accused be arrested and brought before the nearest magistrate.[5]

3. *Execution of the warrant.* To execute the warrant, a marshal or authorized officer arrests the defendant, who must be brought without unnecessary delay before the nearest federal magistrate.[6]

4. *Initial appearance.* If the offense is not triable by the United States magistrate, as in the case of the felony assumed here, the magistrate informs the defendant of the complaint against him, of his right to counsel, of his right to make a statement that, if made, could be used against him, and of the circumstances under which he may secure pretrial release. The magistrate also informs the defendant of his right to a preliminary examination. At the time of his initial appearance, the defendant is not obliged to plead. It is assumed here that the magistrate has admitted the defendant to bail.[7]

5. *Preliminary examination.* The law provides that the preliminary examination shall be held *not later than 10 days following the initial appearance, if the defendant is in custody, nor later than*

4. FED. R. CRIM. P. 3.
5. Ibid., 4.
6. Ibid., 4 (d).
7. Ibid., 5.

20 days if he is not in custody. Extension of these limits without the defendant's consent is allowed only by showing "that extraordinary circumstances exist and that delay is indispensable to the interests of justice." Even with the defendant's consent, the limits cannot be extended without a magistrate's finding of good cause for extension, "taking into account the public interest in the prompt disposition of criminal cases."[8]

If the defendant has been indicted by a grand jury, or if he has waived his right to a preliminary examination, none is held. In this case it is assumed that there has been neither indictment nor waiver.

At the preliminary examination a finding of probable cause may be based upon hearsay evidence in whole or in part. The defendant is entitled to cross-examine and to offer evidence in his own behalf, but he may not object to evidence on the ground that it was acquired by unlawful means; such motions to suppress must be made to the trial court.[9]

6. *Indictment.* A finding of probable cause at the preliminary examination allows the complaint to continue as a charging document while the prosecuting attorney brings the complaint before the grand jury, which must decide whether or not to indict the defendant. *The indictment must be returned within 30 days of arrest, or the complaint is dismissed.*[10]

It is assumed here that an indictment is returned, a decision that obligates the prosecuting attorney to make public the finding and, more pertinently, *to set a trial date not later than 70 days from the date of the indictment.*[11]

7. *Arraignment.* It is also the obligation of the prosecuting attorney to schedule an arraignment as soon as possible after the indictment so that the trial can be held comfortably within the 70-day period. At the arraignment the defendant is obliged to plead; in this case a plea of not guilty is assumed.

8. *Pretrial conference.* After the indictment, *within the 70-day limitation*, a pretrial conference involving judge, prosecutor, and defense counsel is held to consider motions that have been filed and to plan for conduct of the trial.

9. *Trial.* It is assumed here that the defendant is tried and found guilty.

10. *Presentence investigation.* Following conviction, an in-

---

8. Ibid., 5 (c).
9. Ibid., 5.1.
10. Speedy Trial Act, 18 U.S.C. §§3161 (b), 3162.
11. Ibid., §3161 (c)(1).

vestigation is made of the defendant's family circumstances, work history, previous criminal record, and other matters related to probable future behavior. These findings are reported to the trial judge.[12]

11. *Sentence.* Based upon the seriousness of the crime for which the defendant has been convicted, the findings of the presentence investigation, and, of course, judgment, sentence is imposed by the trial judge.

12. *Appeal, incarceration.* Having been sentenced, the defendant is allowed 10 days in which to file an appeal.[13] It is assumed here that an appeal is not made, so the sequence ends with incarceration.

There are many possible variations of this 12-operation sequence. In the context of exemplifying jobbing system technology, however, this example is eloquent.

In jobbing systems, as has been said, control of key operations is necessary if inordinate delays are to be avoided. Prior to passage of the Speedy Trial Act, sequences of the kind described here were several times longer than the intervals now prescribed by law. Evidence of this is the fact that in order to accommodate conditions existing at the time the new law became effective, the act permitted an interval of 180 days between arraignment and trial for the first 12-month period, 120 days for the second, and 80 days for the third. Phase-in to a 60-day limitation between arraignment and trial did not take effect until the beginning of the fourth year.[14] The current 70-day limitation between indictment and trial (rather than 10 days between indictment and arraignment, and 60 days between arraignment and trial) took effect in 1979.[15]

To provide necessary flexibility, the law specifies a number of permissible causes for delay but does so in a tone that connotes urgency and, in the case of the courts' calendars, does not condone procrastination.[16] There are also sanctions against delay, among them possible dismissal of the complaint against the defendant.[17]

Jobbing system performance cannot be improved just by passing a law; if a system is saturated, lengthy delays cannot be reduced by edict. There are, however, remedies. One is to increase system capacity; in the federal system this was done by appointing additional

---

12. Fed. R. Crim. P. 32 (c).
13. Ibid., 4 (b).
14. Speedy Trial Act, §3161 (g).
15. Act of August 2, 1979, Pub. L. No. 96-43, 2.
16. Speedy Trial Act, §3161 (h)(8)(C).
17. Ibid., §3162 (a)(1).

judges. A second measure has been to assign priorities; in the federal system criminal cases have been given priority over civil cases. In the manner of precedence given to emergency over elective surgery in hospitals, civil cases can be deferred with less chance of incurring injustice. A third measure, palliative during peak loads, has been to call upon the services of semiretired judges, a practice that is, in essence, a form of subcontracting.

It is possible that legal and judicial processes are the most difficult of all jobbing systems to control. Despite this, the Speedy Trial Act demonstrates that control flexibly applied can serve to exalt justice, often so long deferred as to be denied.

NUCLEAR POWER

A final jobbing system example is not intended to demonstrate the values of eliminating unnecessary operations, diminishing division of labor, or gaining control over key work centers. It is intended only to emphasize an economic fact: that delays inherent in jobbing sequences are more costly than is commonly realized.

Table 5.1 shows but one part of a regulatory procedure required to secure from the Nuclear Regulatory Commission a construction permit for a nuclear power plant. The sequence extends over more than a year. Similar procedures and the processes of construction, both aggravated and extended by legal and social interventions, have lengthened to more than a decade—from ten to twelve years—the interval between resource input and sale of electricity from the new facility.

During this time the utility receives no revenue from the unfinished facility but must continue to supply capital for it. There are two internal sources for this capital: the utility may use such funds as are available from accumulated depreciation, and it may apply some portion of retained earnings. But for most of the funds needed—considerably more than half—the utility must seek funds from investors.

Buyers of stocks and bonds from regulated utilities expect yield in preference to growth, and this expectation must be met by payments of dividends and interest during the entire period of construction, while there is zero revenue from the new plant. These monetary costs are therefore proportional, as pointed out in equation 1.5, to the interval between commencement of resource input and the beginning of generation and sale of electricity.

For the decade-long regulatory and construction process, the

Table 5.1 Intervals for securing a construction permit for a nuclear power plant

Time (in days)	Operational step
$T_0–T_{30}$	Acceptance review of the utility's environmental report
$T_{30}–T_{60}$	Site visit by NRC's interdisciplinary team
$T_{60}–T_{67}$	Scoping the assessment
$T_{67}–T_{110}$	Preliminary draft environmental statement
$T_{110}–T_{140}$	Publication of NRC's draft environmental statement (DES)
$T_{140}–T_{185}$	Review of DES by agencies, intervenors, etc.
$T_{185}–T_{215}$	Publication of NRC's final environmental statement (FES)
$T_{215}–T_{245}$	Review of FES by Council on Environmental Quality
$T_{245}–T_{275}$	Preconference hearing by Atomic Safety Licensing Board (ASLB)
$T_{275}–T_{310}$	ASLB public hearings
$T_{310}–T_{340}$	ASLB findings of fact
$T_{340}–T_{370}$	ASLB recommendations to NRC
$T_{370}–T_{380}$	NRC grants construction permit (CP)

SOURCE: *Panel on Public Policy on Nuclear Energy for Electricity Generation*, Final Report to the Governor of Maryland, College Park, Md., prepared by the Center for Environmental and Estuarine Studies, University of Maryland, 1976, p. 60. The data were provided by Edward G. Struxness, Associate Director, Environmental Sciences Division, Oak Ridge National Laboratory, Oak Ridge, Tenn.

cost of money is about half as great as the cost of construction itself and about one-third the total cost of the power plant.[18]

Who will pay these time-dependent costs of money?

Those who use the electricity generated by the facility's reactor. And the longer the interval, the more users will pay.

Utility rates are determined by an allowed rate of return, expressed as a percentage, applied to the depreciated value of the plant. Included in this value is the cost of money, made excessively high by the many steps and delays of regulation, delays made longer by the objections of interveners.

Behind this process lies a tacit assumption: that time enhances safety. Safety is not time dependent; cost is.

18. *Panel on Public Policy on Nuclear Energy for Electricity Generation*, Final Report to the Governor of Maryland, College Park, Md., prepared by the Center for Environmental and Estuarine Studies, University of Maryland, 1976, p. 63 (from a presentation to the panel by James L. Everitt, President, Philadelphia Electric Company).

SIX

Articulated Systems

At some long ago time some unknown manager of a jobbing system may have perceived that within the heterogeneity of customers' demands there were patterns of preference for the same products or the same services. Perception may then have led to conception: that repetition could make possible a new kind of system, designed for *sameness* rather than *variety*.

By providing for the performance of but one task, such a concept permits the arrangement of operations in fixed sequence, the juxtaposition of work centers and their component work stations, and the approximate balancing of operation time requirements. These attributes of sameness—sequencing, juxtaposing, and balancing—describe operations technologies that are articulated systems.[1]

SYSTEM CHARACTERISTICS

Such systems, designed to do but one thing, offer great advantages over their jobbing system predecessors, but there are prerequisites to their realization:

All systems, however capable or incapable of variety, depend upon demands for their goods and services, upon the markets

---

1. When first conceptualized, articulated systems were called *integrated*, and they were so characterized in a previous publication (Robert H. Roy, *The Cultures of Management* [Baltimore: Johns Hopkins University Press, 1977], chap. 10). In preparing the present book, I realized that the word *integrated* has been preempted to define systems enlarged by extension toward sources of raw materials or toward markets. In this sense, all kinds of system technologies can be integrated, without, at the same time, being articulated into a fixed, juxtaposed, and approximately balanced sequence of component operations.

they exist to serve. If there is but one market, served by but one product or service, the dependence of system upon market becomes a form of commensalism, a word that describes the dependence of a living organism upon a single source of food, like the dependence of the koala upon the leaves of eucalyptus, or the panda upon those of bamboo. In this sense, single-purpose articulated system technologies are like commensable species in nature: more vulnerable than systems attuned to variety. Should demand be fickle or ephemeral, and diminish, cease, or change, articulated systems could quickly become obsolete.

Articulated systems require "prenatal" design; they cannot "just grow" in the manner of their jobbing system predecessors. The potential market and its prospective longevity must be examined in order to estimate revenues for comparison with the capital required to equip the system and sustain its operations. Fixed capital must be invested in advance, and if special-purpose machinery and equipment are to be procured, fixed-capital needs are likely to be greater for articulated systems than for jobbing systems. Because of the need for "up-front" fixed capital, dependence upon market continuity, and use of special-purpose equipment, investment in articulated system technology incurs greater financial risk.

Articulated system design requires identifying the sequence of work centers, choosing the machinery, equipment, and personnel to be used, and estimating the time that will be taken at each work station to process a unit of product or service. These data will determine the number of work centers and their component work stations in proportion to that operation's unit time interval. Thus, if a sequence of five operations requires 1, 3, 2, 6, and 3 time units per unit of product or service, there must be one work station for Work Center 1, three work stations for Work Center 2, two for Work Center 3, six for Work Center 4, and three for Work Center 5, arranged as shown in Fig. 6.1.

Whole-number proportions of the kind assumed here are very unlikely, and when balance cannot be economically approximated by calculation of a least-common multiple, some idleness or overtime must be accepted at one or more work centers. If, for example, requirements for Work Center 4 were 5.5 time units instead of 6, the system designer would

Fig. 6.1 Operation process chart of the work stations required in a five-center sequence. Work station 4-7, indicated by dotted lines, is related to the discussion of Table 7.1.

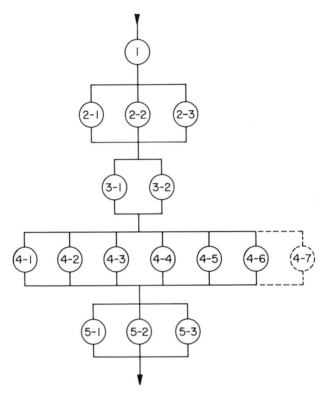

have two options: (1) he could multiply all time units by two and design for 2, 6, 4, 11, and 6 work stations at the five work centers to achieve balance; or (2) he could design for six work stations at Work Center 4 and accept the fact that as a group the six stations would be idle 10 percent of the time.

These alternatives are oversimplified, but they illustrate the conflicting values that the designer must weigh. Balance achieved by using a least-common multiple increases system capacity and fixed-capital requirements by a factor equal to the multiplier. Any lesser multiplier sacrifices balance and incurs the cost of idleness while also reducing system capacity and capital investment. What has been called the "prenatal" design of articulated systems is by no means easy, but the rewards of successful design, as will be shown, are worth striving for.

Once started, operational continuity becomes much more important in an articulated system than in jobbing. If a work center ceases to function in a jobbing system, other parts of the system, at least for a time, can "feed upon" their considerable in-process inventories and their capabilities for variety. But if any work center in an articulated system fails, the entire system must soon cease to operate, there being little in-process inventory or system capability to adapt to other products or services.

To guard against shut-downs, some redundancies must be included in the design of articulated systems: reserves of raw materials, equipment, and personnel, to be called upon when deliveries are late, when machines break down, or when workers are absent. In the design process, the capital and operating costs of redundancy must be weighed against the benefits expected from avoidance of interruptions.

Offsetting these requisites of articulation are advantages that ameliorate the constraints that attend the creation and operation of articulated systems:

Demands at points of entry into articulated systems are determined by system managers, not by the uncontrolled volition of individual customers and clients. Management decisions are based upon market forecasts and assessments of inventories of raw materials and finished goods. On occasions of slack demand, the manager of an articulated system may elect to produce for the inventory of finished goods, an option denied to jobbing system managers. As a consequence of these buffers, articulated systems enjoy more uniform input and greater operational stability than jobbing systems.

The combination of invariable operation sequences and approximate balance between centers permits close spacing between component tasks. Moving times, M_j^i, and moving distances therefore are small.

Approximate equality between operation times at work stations $(P_1^i \approx P_2^i \approx P_3^i \approx \ldots P_j^i)$ makes possible very small inventories of work in process, V, and less space, A, is required for them.

As a consequence of approximate balance among and between work centers, interoperation delays, $_jD^i$ and D^i_j, are reduced. In jobbing systems, delays exact penalties deriving from division of labor as well as from inventory cushions against variety. In articulated systems, because of small in-process inventories and brief delays, division of labor does not exact such penalties.

Single-product repetition at each work station reduces performance times, P^i_j. All operations, of whatever kind, may be divided into three components: make-ready, do, and put away. The first and last of these may be regarded as constants, not dependent upon the magnitude of the "do" component. The duration of that component will be proportional to the number of units to be processed.[2] In articulated systems, where each succeeding unit of product or service is just like its predecessor, make-ready and put away are eliminated and the facility of "doing" is enhanced.

Short intervals for moves and delays and shorter intervals for processing result in much quicker job progression through articulated systems. Relative to jobbing systems, articulated system technology provides much lower values for T^i and C^{i*}

Control of articulated systems is embodied in each system's design. It is necessary only to dispatch jobs to the first work center, from which they will progress in fixed sequence and approximate balance to each successor work center, emerging from the final station completed. To correct imbalances that occur, system managers must increase or decrease work-center production by using overtime or short time, or by employing such redundant resources as are available. On occasions when these remedies do not serve—when, for example, there is a major breakdown—the system as a whole must stop.

---

2. To say that processing time is proportional to the number of units produced is true enough for the point made, but this is not quite the whole truth. There are "warm-up" periods at the beginning of the work day, after coffee breaks, and after lunch periods, and sometimes a slackening of productivity before quitting time. These variations occur in many kinds of repetitive tasks but are more significant in jobbing systems, where each different job requires a "learning" period in which to come up to speed.

To illustrate the effects of articulation, Table 2.1 will again be used. In that tabulation, processing intervals for Job No. 1 were assumed to repeat a total of eight times in the 20-job sequence: P^1, P^4, P^7, P^8, P^{11}, P^{13}, P^{16}, and P^{19}.

Imagine that the manager of that assumed jobbing system observed that this processing sequence of

$$P_1 = 5 \text{ minutes}$$
$$P_2 = 20 \text{ minutes}$$
$$P_3 = 12 \text{ minutes}$$
$$P_4 = 40 \text{ minutes}$$
$$P_5 = 17 \text{ minutes}$$

occurred again and again, often enough for him to conclude that the market for this particular good or service was sufficient to warrant articulation.

Assume also that design, time measurements, and pilot trials led to conclusions that the five component operations could be juxtaposed in fixed sequence, that each operator would immediately move completed work to the next station, and that processing intervals and moving times would be as follows:

$$M_1 + P_1 + M_2 = 1 + 4 + 1 = 6 \text{ minutes}$$
$$P_2 + M_3 = 19 + 1 = 20 \text{ minutes}$$
$$P_3 + M_4 = 11 + 1 = 12 \text{ minutes}$$
$$P_4 + M_5 = 38 + 2 = 40 \text{ minutes}$$
$$P_5 + M_6 = 16 + 2 = 18 \text{ minutes}$$

Using these data, the manager then plans for a single station at Work Center P_1, for three work stations at Work Center P_2, for two at Work Center 3, seven at work Center 4, and three at Work Center 5. These are arranged as shown in Fig. 6.1, except that an additional work station has been added to Work Center P_4.

The daily capacities at each of these work centers during each 8-hour (480-minute) day are:

$$P_1 = 480/6 = 80 \text{ units}$$
$$P_2 = 3 \times 480/20 = 72 \text{ units}$$
$$P_3 = 2 \times 480/12 = 80 \text{ units}$$
$$P_4 = 7 \times 480/40 = 84 \text{ units}$$
$$P_5 = 3 \times 480/18 = 80 \text{ units}$$

By these assumptions, the five-stage system shows three work centers, 1, 3, and 5, in balance, each capable of producing 80 units per day; another, Work Center 2, capable of turning out only 72 units per day; and Work Center 4, with a higher capacity of 84.

If system output is determined by the slowest work center and is set at 72, the other four work centers must suffer intervals of idleness or work fewer hours. If output is pegged at 84 units per day, all four of the lower-capacity centers must work overtime to keep up. If the three centers that are in balance determine that output shall be 80 units per day, Work Center 2 must produce more by working either faster or longer, and Work Center 4 must work for a shorter time and/or have intermittent periods of idleness.

One of many possible decisions, not necessarily the best nor by any means the only one, is shown in Table 6.1. In the interest of simplicity, the sequence unrealistically disregards variances and more reasonably assumes that the cost of interunit sequencing idleness, $_iI$, is greater than the carrying cost of in-process inventory. In no case can work begin on any unit until that unit has been completed and moved from a preceding work center. If completion and move from a predecessor center take place before completion of a unit in process, there will be a preoperation delay, $_iD$, at the successor center, as shown in the delay columns for all four work centers. If completion and move do not take place until after the next station completes its work upon the preceding unit, there will be work center idleness $_2I$, as shown for Units 10, 11, and 12 at Work Center 2.

The tabulation assumes that Work Center 1 has worked for a preliminary hour to produce Units 1–9, to create an in-process inventory that Work Center 2 will begin processing the following day.

Each day thereafter Work Center 2 will commence work at time zero and Work Centers 1 and 3 will begin work one hour later, at $T = 60$.

Work Center 1 will finish the day at $T = 540$, a working interval of 8.0 hours. During this time, Units 10–89 will have been completed, without idleness or delay. Units 81–89 will be held to start the following day at Work Center 2.

The three work stations at Work Center 2 start earlier, at $T = 0$, and work for approximately 9 hours, finishing at $T = 546$, $T = 552$, and $T = 538$. During the day, Units 1–80 have been completed.

Disregarding the overnight layover of Units 1–9, six of these previously completed units (Units 4–9) will incur preoperation delays $_2D$, and so, beginning with Unit 13, will succeeding units produced at Work Center 1, when work begins there at $T = 60$. The sum of these delay times is 1,790 minutes, an average in-process delay $_2\overline{D}$ of 22.4 minutes per unit. Because of the assumption that Work Center 1 starts at $T = 60$, and because units processed at Work Center 2 cannot start until Units 10, 11, and 12 become available, the three work stations at Work Center 2 show idleness $_2I$ for a total of 36 minutes.

The two work stations at Work Center 3 start at $T = 60$ and work until $T = 558$ and $T = 564$, intervals of 8.3 and 8.4 hours, respectively. Units 1–80 are completed. Because of the delayed start relative to Work Center 2, there are preoperation delays $_3D$ aggregating 882 minutes, ranging from a high of 52 minutes for Unit 3 to lows of zero later in the day. The average in-process delay $_3\overline{D}$ is $882/80 = 11.0$ minutes per unit. Short intervals of sequencing idleness $_3I$ total 42 minutes.

At the fastest-producing center, Work Center 4, a later start at $T = 108$ has been assumed. At this seven-station work center, stopping times vary: at $T = 590$ for work station 4-1, $T = 598$ for 4-2, $T = 604$ for 4-3, $T = 564$ for 4-4, $T = 570$ for 4-5, $T = 578$ for 4-6, and $T = 584$ for 4-7. This gives a range for the work day of from 7.6 hours to just under 8.3 hours, about which more will be said later. Units 1–80 are completed. Idle time $_4I$ totals 132 minutes and preoperation delay time $_4D = 876$ minutes, an average of 11.0 minutes per unit.

The three work stations of Work Center 5 start at $T = 148$, with Work Station 1 finishing at $T = 634$, Work Station 2 at $T = 638$, and Work Station 3 at $T = 624$, a working-day range of from 7.9 hours to a little less than 8.2 hours. Preoperation delays $_5D$ total 1,704 minutes, an average of 21.3 minutes per unit. Idle time $_5I$ for the three stations is 12 minutes.

On succeeding days the cycle repeats. Work Center 1, again starting at $T = 60$, produces Units 90–169, Work Center 2 starts at $T = 0$, Work Center 3 at $T = 60$, Work Center 4 at $T = 108$, and Work Center 5 at $T = 148$. Work Centers 2, 3, 4, and 5 all complete Units 81–160. Units 161–69 from Work Center 1 are held over for the following day.

These data appear to be contrived, and indeed they are, but they allow me to make several points about the management of articulated systems:

Work centers in articulated systems are interdependent. Malfunction at one center spreads wavelike, forward through or backward into other centers, and sometimes managerial intervention is required to restore balance.

Table 6.1 Hypothetical flow of work through a five-operation articulated system

Work Center (time in minutes)

Unit number	1 Start	1 Finish	2 Station	2 Arrive	2 Start	2 Finish	2 Idle	2 Delay	3 Station	3 Arrive	3 Start	3 Finish	3 Idle	3 Delay	4 Station	4 Arrive	4 Start	4 Finish	4 Idle	4 Delay	5 Station	5 Arrive	5 Start	5 Finish	5 Idle	5 Delay
1			1	0	**0**	20	0	0	1	20	**60**	72	0	40	1	72	**108**	148	0	36	1	148	**148**	166	0	0
2			2	0	**0**	20	0	0	2	20	**60**	72	0	40	2	72	**108**	148	0	36	2	148	**148**	166	0	0
3			3	0	**0**	20	0	0	1	20	72	84	0	52	3	84	**108**	148	0	24	3	148	**148**	166	0	0
4			1	0	20	40	0	20	2	40	72	84	0	32	4	84	**108**	148	0	24	1	148	166	184	0	18
5			2	0	20	40	0	20	1	40	84	96	0	44	5	96	**108**	148	0	12	2	148	166	184	0	18
6			3	0	20	40	0	20	2	40	84	96	0	44	6	96	**108**	148	0	12	3	148	166	184	0	18
7			1	0	40	60	0	40	1	60	96	108	0	36	7	108	**108**	148	0	0	1	148	184	202	0	36
8			2	0	40	60	0	40	2	60	96	108	0	36	1	108	148	188	0	40	2	188	188	206	4	0
9			3	0	40	60	0	40	1	60	108	120	0	48	2	120	148	188	0	28	3	188	188	206	4	0
10	**60**	66	1	66	66	86	6	0	2	86	108	120	0	22	3	120	148	188	0	28	1	188	202	220	0	14
11	66	72	2	72	72	92	12	0	1	92	120	132	0	28	4	132	148	188	0	16	2	188	206	224	0	18
12	72	78	3	78	78	98	18	0	2	98	120	132	0	22	5	132	148	188	0	16	3	188	206	224	0	18
13	78	84	1	84	86	106	0	2	1	106	132	144	0	26	6	144	148	188	0	4	1	188	220	238	0	32
14	84	90	2	90	92	112	0	2	2	112	132	144	0	20	7	144	148	188	0	4	2	188	224	242	0	36
15	90	96	3	96	98	118	0	2	1	118	144	156	0	26	1	156	188	228	0	32	3	228	228	246	4	0
16	96	102	1	102	106	126	0	4	2	126	144	156	0	18	2	156	188	228	0	32	1	228	238	256	0	10
17	102	108	2	108	112	132	0	4	1	132	156	168	0	24	3	168	188	228	0	20	2	228	242	260	0	14
18	108	114	3	114	118	138	0	4	2	138	156	168	0	18	4	168	188	228	0	20	3	228	246	264	0	18
19	114	120	1	120	126	146	0	6	1	146	168	180	0	22	5	180	188	228	0	8	1	228	256	274	0	28
20	120	126	2	126	132	152	0	6	2	152	168	180	0	16	6	180	188	228	0	8	2	228	260	278	0	32
21	126	132	3	132	138	158	0	6	1	158	180	192	0	22	7	192	192	232	4	0	3	232	264	282	0	32
22	132	138	1	138	146	166	0	8	2	166	180	192	0	14	1	192	228	268	0	36	1	268	274	292	0	6
23	138	144	2	144	152	172	0	8	1	172	192	204	0	20	2	204	228	268	0	24	2	268	278	296	0	10
24	144	150	3	150	158	178	0	8	2	178	192	204	0	14	3	204	228	268	0	24	3	268	282	300	0	14
25	150	156	1	156	166	186	0	10	1	186	204	216	0	18	4	216	228	268	0	12	1	268	292	310	0	24
26	156	162	2	162	172	192	0	10	2	192	204	216	0	12	5	216	228	268	0	12	2	268	296	314	0	28
27	162	168	3	168	178	198	0	10	1	198	216	228	0	18	6	228	228	268	0	0	3	268	300	318	0	32
28	168	174	1	174	186	206	0	12	2	206	216	228	0	10	7	228	232	272	0	4	1	272	310	328	0	38
29	174	180	2	180	192	212	0	12	1	212	228	240	0	16	1	228	268	308	0	40	2	308	314	332	0	6
30	180	186	3	186	198	218	0	12	2	218	228	240	0	10	2	240	268	308	0	28	3	308	318	336	0	10
31	186	192	1	192	206	226	0	14	1	226	240	252	0	14	3	240	268	308	0	28	1	308	328	346	0	20
32	192	198	2	198	212	232	0	14	2	232	240	252	0	8	4	252	268	308	0	16	2	308	332	350	0	24
33	198	204	3	204	218	238	0	14	1	238	252	264	0	14	5	252	268	308	0	16	3	308	336	354	0	28
34	204	210	1	210	226	246	0	16	2	246	252	264	0	6	6	264	268	308	0	4	1	308	346	364	0	38
35	210	216	2	216	232	252	0	16	1	252	264	276	0	12	7	276	276	316	4	0	2	316	350	368	0	34
36	216	222	3	222	238	258	0	16	2	258	264	276	0	6	1	276	308	348	0	32	3	348	354	372	0	6
37	222	228	1	228	246	266	0	18	1	266	276	288	0	10	2	288	308	348	0	20	1	348	364	382	0	16
38	228	234	2	234	252	272	0	18	2	272	276	288	0	4	3	288	308	348	0	20	2	348	368	386	0	20
39	234	240	3	240	258	278	0	18	1	278	288	300	0	10	4	300	308	348	0	8	3	348	372	390	0	24
40	240	246	1	246	266	286	0	20	2	286	288	300	0	2	5	300	308	348	0	8	1	348	382	400	0	34
41	246	252	2	252	272	292	0	20	1	292	300	312	0	8	6	312	312	352	4	0	2	352	386	404	0	34
42	252	258	3	258	278	298	0	20	2	298	300	312	0	2	7	312	316	356	0	4	3	356	390	408	0	34

Bottom table (continuation):

Unit																	
43	258	264	1	264	286	306	0	22	306	306	1	306	312	324	0	0	324	6
44	264	270	2	270	292	312	0	22	312	312	2	312	318	324	2	0	324	0
45	270	276	1	276	298	318	0	22	318	318	1	318	324	336	0	2	336	6
46	276	282	2	282	306	326	2	24	326	324	3	324	330	348	0	0	348	0
47	282	288	2	288	312	332	0	24	332	336	1	336	348	348	0	0	348	4
48	288	294	2	294	318	338	0	24	338	342	2	342	352	350	2	0	352	4
49	294	300	1	300	326	346	0	26	346	348	1	348	360	360	1	0	360	2
50	300	306	1	306	332	352	2	26	352	354	2	354	366	364	2	0	364	2
51	306	312	3	312	338	358	2	26	358	358	3	358	360	372	1	2	372	2
52	312	318	1	318	346	366	0	28	366	366	1	366	366	378	2	0	378	0
53	318	324	2	324	352	372	2	28	372	372	2	372	372	384	1	2	384	0
54	324	330	3	330	358	378	0	28	378	386	3	384	390	390	2	2	390	0
55	330	336	1	336	366	386	2	30	386	392	1	390	398	398	1	4	398	0
56	336	342	2	342	372	392	2	30	392	398	2	398	404	404	2	4	404	0
57	342	348	1	348	378	398	2	30	398	406	1	404	410	410	1	0	410	2
58	348	354	2	354	386	406	2	30	406	414	2	410	418	418	2	0	418	2
59	354	360	1	360	392	412	2	32	412	420	1	418	424	424	1	0	424	2
60	360	366	2	366	398	418	2	32	418	426	4	424	430	430	3	2	430	2
61	366	372	1	372	406	426	2	34	426	432	5	438	438	438	1	8	438	0
62	372	378	2	378	412	432	2	34	432	438	6	444	444	444	2	6	444	0
63	378	384	3	384	418	438	2	34	438	444	7	450	450	450	3	6	450	0
64	384	390	1	390	426	446	2	36	446	450	1	458	468	458	1	0	458	0
65	390	396	2	396	432	452	2	36	452	458	2	464	468	464	3	4	464	0
66	396	402	1	402	438	458	2	36	458	466	3	470	470	470	2	0	470	0
67	402	408	2	408	446	466	2	38	466	472	4	478	478	478	3	8	478	0
68	408	414	1	414	452	472	2	38	472	478	5	484	484	484	2	6	484	0
69	414	420	3	420	458	478	1	38	478	484	6	490	490	490	1	6	490	0
70	420	426	2	426	466	486	2	40	486	490	7	498	498	498	3	8	498	0
71	426	432	2	432	472	492	0	40	492	498	1	504	504	504	2	0	504	0
72	432	438	3	438	478	498	2	40	498	506	2	510	510	510	3	2	510	0
73	438	444	1	444	486	506	0	42	506	512	3	518	518	518	1	8	518	0
74	444	450	2	450	492	512	2	42	512	518	4	524	524	524	4	6	524	0
75	450	456	3	456	498	518	2	42	518	526	5	530	530	530	2	6	530	0
76	456	462	1	462	506	526	2	44	526	532	6	538	538	538	3	6	538	0
77	462	468	2	468	512	532	2	44	532	538	7	544	544	544	1	6	544	0
78	468	474	1	474	518	538	2	44	538	550	1	550	550	550	3	2	550	0
79	474	480	3	480	526	546	0	46	546	550	2	558	558	558	2	8	558	0
80	480	486	2	486	532	552	2	46	552	558	3	564	564	564	1	6	564	0
81	486	492																
82	492	498																
83	498	504																
84	504	510																
85	510	516																
86	516	522																
87	522	528																
88	528	534																
89	534	540																
Totals			36					1,790							42	882		

NOTE. There is no idleness or delay at Work Center 1. Units 1–9 have been in storage since the preliminary period, and units 81–89, 161–169, etc., will be in overnight storage on succeeding work days. Shut-down periods have not been included in delay intervals.

65

Flexibility in working hours is important to the maintenance and restoration of balance. Acceptance of different starting and stopping times requires the consent of employees as well as ordination by management. In this sense the assumptions of Table 6.1 may presume too much. Multishift, around-the-clock operations confound the difficulties of restoring balance by making changes in working hours.

Variances in times required for processing, moves, and delays put additional burdens upon articulated system managers, but occasionally—and somewhat paradoxically—performance variances can restore stability. Workers, perceiving a bottleneck, may speed up, or, sensing slack, slow down, by their voluntary variations in performance thereby restoring balance.

Among the four classes of elemental system components (moves, processes, delays, and idle intervals), processing and idle times and costs usually command the most attention from management. In the case assumed here, moves are made by production workers, and moving intervals are therefore considered to be in the same category as processing.

System articulation contributes to reductions in all three of these elemental components: processing times are reduced by enhanced celerity from sameness and repetition and by elimination of the changeovers demanded by variety, moving times are reduced by juxtaposition of work stations, and delays are dramatically reduced by approximate balance between work centers.

SYSTEM COMPARISONS

From the data assumed for the values of M and P compiled in Table 6.1, average time values for a single unit of product or service will be as shown in the flow data of Table 6.2. Summaries from the table can now be compared with the jobbing system characteristics computed in chapter 2.

Total time and cycle time

As in the single-channel case, all paths through the assumed articulated system are alike, so that total time and cycle time (equations [1.1] and [1.2]) are equal. For the jobbing system, T^1 and C^{1*} totaled 916 minutes. In the articulated case, using Unit 1 as representative of all the units processed,

Table 6.2　Intervals in minutes for Unit 1, produced by the
assumed articulated system

Element	Time (in min.)	Element	Time (in min.)
M_1	1	P_3	11
$_1D$	0	D_3	0
$_1I$	0	M_4	1
P_1	4	$_4\overline{D}$	876/80 = 11.0
M_2	1	$_4\overline{I}$	132/80 = 1.7
$_2\overline{D}$	1,790/80 = 22.4	M_5	2
$_2\overline{I}$	36/80 = 0.5	$_5\overline{D}$	1,704/80 = 21.3
P_2	19	$_5\overline{I}$	12/80 = 0.2
D_2	0	P_5	16
M_3	1	D_5	0
$_3\overline{D}$	882/80 = 11.2	M_6	2
$_3\overline{I}$	42/80 = 0.5		

Summary:

$$\Sigma M_j \;=\; 1 + 1 + 1 + 1 + 2 + 2 = 8 \text{ min.}$$
$$\Sigma_j\overline{D} \;=\; 0 + 22.4 + 11.0 + 11.0 + 21.3 = 65.7 \text{ min.}$$
$$\Sigma_j\overline{I} \;=\; 0 + 0.5 + 0.5 + 1.7 + 0.2 = 2.9 \text{ min.}$$
$$\Sigma P_j \;=\; 4 + 19 + 11 + 38 + 16 = 88 \text{ min.}$$
$$\Sigma D_j \;=\; 0$$

$$T^1 = C^{1*} = 8 + 66 + 88 + 0 = 162 \text{ minutes.}$$

Idle time, $_jI$, is not included in the summation for total time
and cycle time, because it represents waiting time by the production
centers and does not extend time for work in process.

Operational efficiency

Operational efficiency (equation [1.3]) for the assumed articulated
system will use the same numerator chosen to represent output
from the jobbing system, 94 standard minutes. The denominator, as
before, will be the sum of moving and processing times, plus the
sum of average idle times per unit:

$$E_p^1 = 94/(8 + 2.9 + 88) = 94/98.9 = 0.950 = 95\%.$$

System efficiency

System efficiency for the jobbing system was calculated to be slightly above 10 percent, for the most part because of assumed values for delays $_jD$ and D_j. In the articulated system, efficiency E_s (equation [1.4]) becomes

$$E_s^1 = 94/(8 + 66 + 2.9 + 88 + 0) = 94/164.9$$

$$= 0.570 = 57\%.$$

Working-capital requirements

In the jobbing system described in Chapter 2, working-capital requirements were shown to be proportional to 916 minutes. In the articulated case discussed here, idle time must be added, as in the denominator above, so that working capital $W_j^1 \sim 164.9$ minutes, less than one-eighth the requirement for the jobbing system assumption.

In-process inventory

In-process inventory in the articulated system is proportional to this same value less idle time, giving a result of 162 minutes, compared to the jobbing system value of 916 minutes.

Floor-space requirements

Equation (1.7) does not properly apply to the articulated system postulated here, because in-process floor space must be provided for whatever may be the maximum need at each work station for the assumed operating conditions.[3]

Somewhat unrealistic, but still suitable for purposes of comparison, is the assumption in Table 6.1 that each unit requires space at Work Center 1 only during the time the product or service is moved to the center, processed there, and moved to Work Center 2. In equation (1.7) these intervals are accounted for by inclusion of M_1^1, P_1^1, and M_2^1 (6 minutes).

3. All forms of operational technologies must provide floor space for whatever maximum requirements are likely to occur, not just for a single unit or batch of product or service, but for the fluctuating demands of all work in the system over time. Equation (1.7) reflects space needs for only a single job, and this provides an adequate measure for the jobbing system, but space needs for the articulated system are determined by sequencing imbalances, so that approximate maxima must be considered in making a comparison.

At Work Center 2 there must be space at the three work stations for Units 1–9, produced on the previous day, and for Units 81–89, and so on, on succeeding days. But more space than this is required. Beginning at $T = 66$, and continuing each 6 minutes thereafter, an additional unit arrives from Work Center 1, while the three work stations of Work Center 2 are completing and moving units to work Center 3 at a rate of one each $6\frac{2}{3}$ minutes (20/3). This leads to maximum delay times for Units 78, 79, and 80 of $44 + 46 + 46 = 136$ minutes. This $_2D^1$ equivalent, plus $P_2^1 + M_3^1 = 20$ minutes, substitutes in equation (1.7).

Analogously, $40 + 52 = 92$ minutes (Units 2 and 3) becomes the space equivalent for Work Center 3; $36 + 36 + 24 + 24 + 12 + 12 + 0 = 144$ minutes (Units 1–7) becomes the equivalent for the seven work stations of Work Center 4; and $34 + 34 + 34 = 102$ minutes (Units 40, 41, and 42), the equivalent for the three work stations of Work Center 5.

Substituting these for $_jD^1$ in equation (1.7), along with times for M_j^1 and P_j^1 $(D_j^1 = 0)$, gives

$$A_j^1 \sim 6 + (20 + 136) + (12 + 92) + (40 + 144) + (18$$

$$+ 102) = 6 + 156 + 104 + 184 + 120 = 570 \text{ minutes.}$$

This equivalent for space requirements in the articulated system is brought about by the center-to-center imbalances assumed for values of M and P. Even so, the 570-minute figure is only about two-thirds as large as the 916-minute equivalent found for Job No. 1 in the jobbing system.

Summary comparisons

Comparisons between the assumed jobbing and articulated systems can be summarized as follows:

		Jobbing	Articulated
Total time	T^1 =	916	162
Cycle time	C^{1*} =	916	162
Operational efficiency	E_p^1 =	100%	95%
System efficiency	E_s^1 =	10%	57%
Working-capital requirements	$W^1 \sim$	916	165
In-process inventory	$V^1 \sim$	916	162
Floor-space requirements	$A^1 \sim$	916	570

These comparisons are biased in favor of articulation by assuming that there are no stoppages $(H = 0)$.

If one of the multiple work stations of Work Centers 2, 3, 4, and 5 were to malfunction, it might be possible to continue system operation by working other work stations overtime. If, however, the single station at Work Center 1 were to fail, the entire system would quickly come to a halt and H would become positive. Depending upon the duration of such stoppages and the frequencies of their occurrence, all the comparisons shown above would become less favorable for articulation.

If malfunctions at Work Center 1 proved to be sufficiently onerous, stoppage disruptions could be ameliorated by adding a second, stand-by work station. The cost associated with such an investment might be less than the cost of single-station stoppages, but the advantages shown for articulation in these comparisons would be diminished.

CONCLUSION

The extent to which these numbers can be called representative, depending as they do upon oversimplified assumptions, is, of course, not known. Nevertheless, the degrees of difference are sufficiently large for us to conclude that the "invention" of articulation, by whatever means and whenever first contrived, was a contribution of great value. Incalculable numbers of mass-produced and mass-marketed goods and services have been and are provided by systems that, wholly or in part, have embraced the principles of sequencing, juxtaposition, and balancing, which are the hallmarks of articulation.

SEVEN

Articulated Systems

THE FLOUR MILL AT RED CLAY CREEK
THE RAIL MILL AT STEELTON
BOOKMAKING

As will be shown in the last of the three examples given in this chapter, conversion from jobbing to articulated system technology is often attended by opportunities to combine contiguous operations in balance, to automate transfers from one production center to another, and, sometimes, to continue processing during transfer intervals. Most articulated systems are therefore hybrids, combining articulated, balanced, continuous, and automated sequences, served logistically by jobbing system support for supply, maintenance, and office procedures. Articulated system intentions usually become hybrid system realizations.

THE FLOUR MILL AT RED CLAY CREEK, DELAWARE

Such realizations were captured as long ago as 1787 in Oliver Evans's patented flour mill (Fig. 7.1).[1]

In the sense that earlier flour mills had produced but one product, they may have been either jobbing or articulated systems. (They were also integrated, converting a basic raw material, grain, into a marketable product, flour.) Oliver Evans's contribution, however, was definitely articulation, accompanied by intercenter balances, a modest amount of continuous processing in the "hopper-boy," and even remarkable evidences of automation. That his patent was marketable is suggested by his broadside of 1787 (Fig. 7.2).

1. Greville and Dorothy Bathe, *Oliver Evans* (Philadelphia: Historical Society of Pennsylvania, 1935).

Fig. 7.1 Oliver Evans's patented milling machinery, 1791. The principal parts of the machinery are identified below. For a complete description of the milling process, see *The Young Mill-wright and Miller's Guide* (1795). Reproduced, by permission, from Greville and Dorothy Bathe, *Oliver Evans* (Philadelphia: Historical Society of Pennsylvania, 1935).

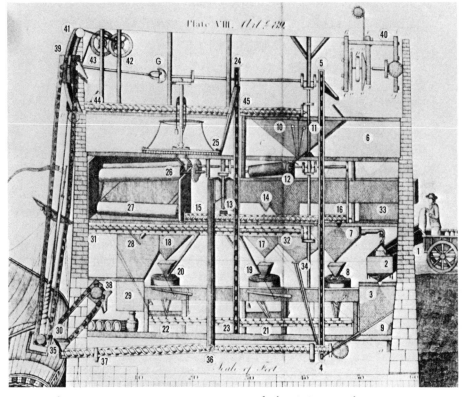

1. The wagoner emptying grain into scale pan.
2. Scale pan, to weigh the grain.
3. Small garner and wind chest for cleaning wheat.
4–5. Elevator, to top floor of mill.
6. Main store of wheat
7. Garner, feeding the shelling or rubbing stones.
8. The rubbing stones.
9. Grain is again elevated, then deposited in garners 10 and 11.
12. Rolling screen.
13. Fan for cleaning the grain.
15–16. The conveyor to garners 7, 17, and 18.
8, 19, 20. Millstones.
20–21. The conveyor, collecting

meal after it is ground.
23–24. Elevator to hopper-boy 25.
25. Hopper-boy, which spreads and cools the meal and supplies the bolting chest.
26, 27. Bolting reels.
28. The chest containing the superfine flour.
29. Spout for filling barrels.
35–39. Elevator for unloading ships. This rises and falls in the curved slots and is driven by the universal coupling at G.
38. A temporary elevator for short lifts.
40. One view of the mechanism (42–43) for hoisting the elevator clear of the ship.

Fig. 7.2 Broadside advertisement of Oliver Evans's first mill, 1787. Reproduced by permission of the Massachusetts Historical Society.

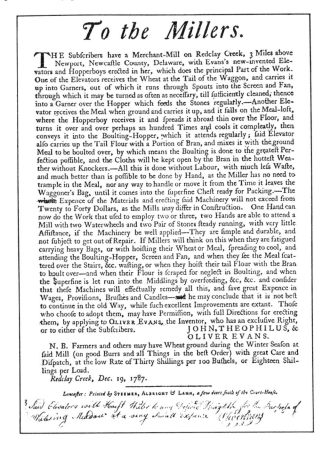

THE RAIL MILL AT STEELTON, PENNSYLVANIA

Another historical example, one that does not evince hybridization, is afforded by the rail mill established in 1866 at Steelton, Pennsylvania.[2]

By that year, refinements in the Bessemer process for converting pig iron to steel had made the operation commercially feasible. Alexander Holley and a group of associates conceived and carried out the design and construction of a mill at Steelton for the single purpose of rolling steel railroad track. The mill was integrated in the sense previously described: it embraced all the processes, from the raw material, iron ore, to the finished product, steel rails.

2. Elting E. Morison, *Men, Machines, and Modern Times* (Cambridge: MIT Press, 1966), chap. 7.

At the time, the market for this single product was growing, as it would for years thereafter. Railroads were rapidly extending their lines throughout the country. Rails made of Bessemer steel were of superior durability and were much preferred over rails of iron.

The Steelton mill was composed of six principal work centers:

Blast furnace: to extract pig iron from iron ore
Converter house: to convert the molten iron to steel by the Bessemer process
Ingot molds: to cool the steel in standard forms
Soaking pits: to heat the ingots to uniform temperature
Blooming mill: to hammer or roll ingots into a shape more suitable for rolling
Rail mill: to roll blooms into finished rails

While it is clear that these processes were integrated, the extent to which they were articulated is not known. Since the operational sequence was fixed and the product uniform, it is reasonable to assume that work centers were arranged in the order listed above, presumably as close together as practicable for the sequence.

Less certain but also probable was system balance. Pig iron, poured from blast furnace extraction, must have been moved quickly, while still in the molten state, to as many Bessemer converters as were necessary to maintain approximate balance. In the converters themselves, the air blown upward through the molten mass burned off the carbon in the pig iron, to effect conversion into steel. The still-molten product then went directly to as many ingot molds as necessary to continue approximate balance among operations. Soaking pits and blooming and rail mills may be assumed to have been provided in numbers suited to this same articulated objective.

These speculations compel belief that there was design awareness of the values of articulation, and they strengthen the conclusion that the system as a whole was, in fact, articulated. Profits from the enterprise remained high for many years, until railway expansion slowed, the replacement market declined, and new technology superseded the Bessemer method.[3]

The Bessemer process did more than make possible the conversion of pig iron to steel; it converted the "iron age" to an "age of steel," and must be regarded as a very great invention indeed. Articulated systems were not created in this same distinctively in-

3. Bessemer converters were superseded by the open-hearth furnace and, more recently, by the oxygen furnace, in which oxygen rather than air is used to burn off unwanted carbon.

ventive way, but the mutations of jobbing system technology into articulated systems have been enormously significant. Articulated technologies are to be found in every kind of manufacturing, and in lesser degree in merchandising, communications, transportation, and systems of government.

BOOKMAKING

For many years the producers of metropolitan newspapers and mass-circulation magazines have used articulated and automated technology for all, or nearly all, system operations, from presswork through distribution. For the most part, the preparatory operations of composition, page makeup, plate making, and so on, have been performed by jobbing systems characterized by deadlines, deadlines met by the provision of stand-by workers, excess equipment, and overtime, accompanied, according to stage, screen, and television, by almost constant frenzy.

Although deadlines exist in every kind of system, those in bookmaking are less stringent and not usually so melodramatic as those involved in the printing of newspapers and magazines. Bookmaking is resistant to articulation, however. Books—take this one, for example—are made by jobbing technologies capable of manufacturing to customers' specifications: capable of composing different type sizes and typefaces in different line and page lengths, printing with different margins in different inks on different kinds, sizes, and weights of paper, and folding, gathering, sewing, and binding in a variety of ways.

Articulation and automation in the manufacture of books began soon after World War II when an imaginative publishing organization combined integration (ownership of a "captive" printing plant), standardization of specifications, and invention of new materials and methods.[4] Later, it automated the entire system, from rolls of paper to finished books.

The books in question were those published for subscribers to

4. The publisher was Doubleday & Company, and the captive printing plant was at Hanover, Pennsylvania. For his assistance and the pleasure of his company I am very much obliged to Mr. Raymond D. Rodman, General Manager of the Hanover plant during the conversion to the articulated and automated system described. Because the distinction between printing and publishing is often misunderstood, it is important to note that printing is *manufacturing*; publishing is *marketing* that which has been manufactured. For a publisher to own a printing plant is a step toward integration, as was Doubleday's ownership of retail bookstores.

book clubs, a large and stable market. It was decided that these books would be produced in a standard format. Typesizes and typefaces could vary according to manuscript length; type page sizes also could vary, but only in limited ways; the trim size of each book was the same; head, front, tail, and bind margins were consistent for a given type page size; and the color (black) and kind of ink and the paper upon which it was printed were seldom changed.

Typesetting, proofreading, and page makeup were done in another plant, and followed by preparation of rubber printing plates, all of one size, grouped in sets of four pages to fold in order of pagination. Affixed to metal sheets by adhesive, these forms were positioned by guide marks that were the same for every book.

At the completion of a run, the finished forms were removed from the press cylinders and new forms were affixed by clamping the metal sheets bearing new rubber plates around the press cylinders. Since dimensions were held constant, adjustments were unnecessary; it was necessary only to unclamp, remove, replace, reclamp, and briefly inspect before starting the next run. Unless paper or ink required replenishment or replacement, presswork on the new forms could begin in a matter of minutes.

The presses themselves were matched to their single purpose: to print on both sides of the paper fed from a reel, in black ink ("perfecting, web-fed, one-color" machines). At the delivery end, each press was designed to cut the printed web into sheets that fed directly into folding machines, balanced to the speed of the press. The output of these two processes, at the rate of about 12,000 per hour, was folded signatures[5] that were stored on skids or pallets until all the signatures comprising the complete book had been printed and folded.

In bookmaking by conventional means, "endpapers"[6] were affixed by paste or glue applied along the folded bind edge to the front

5. In books of the kind described here, signatures may be defined as sheets of paper with 16 pages printed on each of the two sides, folded in such a way that the pages are in numerical order and the folded signatures are equal in size to the untrimmed width and length of the book. By way of example, pages 1, 4, 5, 8, 9, 12, 13, 16, 17, 20, 21, 24, 25, 28, 29, and 32 would be printed on the "white" side of such sheets, and pages 2, 3, 6, 7, 10, 11, 14, 15, 18, 19, 22, 23, 26, 27, 30, and 31 on the "backup" side. Successive signatures would contain pages 33–64, 65–96, 97–128, etc. When gathered with endpapers, each consisting of four pages, before the first and after the last signature, the result would be a complete book.

6. As suggested above, endpapers appear at the front and back of each book and usually are of stronger and heavier paper than that used for text pages. In conventional bookmaking, end sheets are "tipped" along the bind edges of the first and last sig-

of the first and the back of the last signature. The assembly (collation) of complete books was then accomplished by a gathering machine,[7] and books were then sewed by placing the center of each signature over the "saddle" of the sewing machine, through which the binding thread passes to hold the book's signatures together.

Handling every signature in this way required so much time as to make interoperation balance impracticable, a bottleneck that was eventually overcome by invention.

In the articulated system, endpapers were not affixed to the first and last signatures but, instead, were gathered at the front and back of each book as two additional four-page signatures. Sewing was eliminated altogether by invention of a flexible and strong adhesive.[8] As collated books emerged from the gathering machine, each was clamped and passed over a rotating blade that cut off the now-assembled bind edge, a procedure that in effect rendered each book a collection of not-yet-single leaves. Still clamped, each bind edge was then roughed over an abrasive wheel to provide an adherent surface. Successive books then entered a binding machine, where each first passed over a wheel that applied the adhesive by which the book was held together. To set the adhesive, books were then conveyed along an oval path over electric "Calrod" heaters for drying.

During early development of the system, gathered and glued books were stacked on skids or pallets and then fed by hand into a "tumbler" three-knife trimmer, after which they were again stacked and, as the next operation, stained along each now-smooth top edge and stacked again. Later, the tumbler trimmer and staining were balanced and articulated to eliminate postgathering and temporary storage. Thus, the sequence of gathering, slicing, roughing, gluing, drying, trimming, and staining was articulated and automated, with continuous processing evident during each book's passage over the Calrod heaters around the periphery of the binding machine.

Rounding, the curving of the book's backbone to fit the case binding, and backing, the application of a gauzelike cloth to the spine of each book for later adhesion to the case binding, were fol-

---

natures with paste as the adhesive. As each book is cased-in, the outside front and back sides of the endpapers are glued to the case to hold the book in place within the binding.

7. For editions printed in small quantities (e.g., 500 copies) books are often gathered by hand.

8. Invention of the strong and flexible adhesive has been attributed to Luis de Florez, a consulting engineer and retired rear admiral who died in 1962 after a distinguished career.

Fig. 7.3 *Left*: Cross-section of the "sandwich" arrangement for building-in a book's front and back hinges. *Right*: The spine—and only the spine—of each book is laid carefully over the raised metal strip, as shown in the configuration of one layer of four books. The front and back of each case must impinge only upon the upper and lower boards of each layer.

In the articulated system, after many trials, building-in was accomplished mechanically, in approximate balance with casing-in. After casing-in, each book was clamped to expose the spine by just the right amount; each front and back was compressed to form the hinges, and the book was held clamped for seconds to allow the adhesive to set.

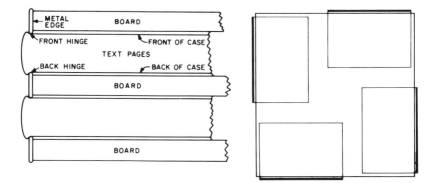

lowed by stacking the rounded and backed books. Rounding was made possible by the flexibility, when heated, of the new adhesive.

In parallel with the operations that have been described, cases were made in a separate sequence of case-making and stamping. Cases and rounded and backed books were then brought together in the casing-in machine, after which each book was piled, sandwich fashion, between layers of rectangular board with four side metal edges raised above and below the top and bottom surface of each board (Fig. 7.3). Stacks of these were then compressed to form the hinge along the front and back of each book, just beside the spine, a process called "building-in." After a delay of approximately 24 hours, to allow the adhesive to set, books were inspected and stacked on skids. Careless stacking caused numerous rejects.

In the articulated system the building-in operation was articulated with casing-in, so that each book entered the now-mechanized building-in process as it emerged from the casing-in machine. This eliminated the careful stacking, compressing, and delay sequences formerly required. Application of jackets, formerly done by hand after inspection, also was mechanized, and inspection changed from examination of every book to quality control by statistical sampling.

Final operations involved insertion into preprepared and pre-labeled cartons, machine sealing, and delivery by a moving belt into the post office, separate from, yet in effect integral with, the plant's system.

Although quantitative data for processing times and intercenter intervals are lacking, it is possible to compare the conventional method of bookmaking with the articulated and automated system by listing the events required for both methods. This has been done in Tables 7.1 and 7.2, each of which is an abridged flow process chart for operations that are common to both technologies. The comparisons are necessarily approximate, but such omissions as there are favor the conventional method, since no account has been taken of changeover and adjustment times that are common to conventional jobbing practices.

The charts assume a book of 320 text pages printed in 32-page forms on the same kind of web-fed perfecting press. The trim size of each book is assumed to be 5½ × 8½ inches, but these specifications do not affect comparison, nor does the equally arbitrary choice of 50,000 copies.

Printing by both methods is assumed to require ten 32-page forms, two of which are printed at the same time, one on the "white" side and the other on the "backup" side. The web therefore emerges from the impression cylinders bearing 64 pages. Unless the paper is thin, 64 pages is too bulky to fold into a single signature, so the web is slit along the middle as it passes toward the press delivery, and is also sliced across into 32-page sheets about 35 × 45 inches in size.

In the chart for the articulated system (Table 7.2), these sheets pass directly into the folding machines that are integral with the press and balanced to its speed. In the conventional system, printed sheets are delivered to skids, and these are moved to the folding machines, which constitute a separate production center, as shown in Table 7.2.

To print a 320-page book in the manner described requires five passes through the press (or one pass through five presses, etc.). Since books do not become single units until gathered, both charts make the simplified assumption that all printing and folding intervals encompass the 5 × 50,000 = 250,000 press impressions on both sides of the paper, and the 10 × 50,000 = 500,000 folded signatures. An analogous assumption for the conventional technology is made

Table 7.1 Abridged flow process chart: Conventional technology

P_1^c	Print 250,000 copies of ten 32-page forms (two up). Slit, slice, and deliver 500,000 sheets to skids.
D_1^c	Await move to folding machines.
M_2^c	Move skids to folding machines.
$_2D^c$	Await folding.
P_2^c	Fold 500,000 sheets into 32-page signatures. Stack on skids.
D_2^c	Await move of first and last signatures to work center for affixing endpapers.
M_3^c	Move first and last signatures to work center for affixing endpapers.
$_3D^c$	Await adhering of endpapers.
P_3^c	Tip 100,000 endpapers to first and last signatures with adhesive. (It is assumed that the operation is mechanized.)
D_3^c	Await move to gathering machine.
M_4^c	Move skids for all signatures to gathering machine.
$_4D^c$	Await gathering.
P_4^c	Gather 50,000 books of ten signatures each. Stack on skids.
D_4^c	Await move to sewing machine work center.
M_5^c	Move skids to sewing machine work center.
$_5D^c$	Await sewing.
P_5^c	Sew $10 \times 50,000 = 500,000$ signatures. Stack on skids.
D_5^c	Await move to three-knife trimmer.
M_6^c	Move skids to three-knife trimmer.
$_6D^c$	Await trimming.
P_6^c	Trim 50,000 books on three sides. Stack on skids.
D_6^c	Await move to staining.
M_7^c	Move skids to staining work center.
$_7D^c$	Await staining.
P_7^c	Stain top edge of 50,000 books. Stack on skids.
D_7^c	Await move to rounding and backing.
M_8^c	Move skids to rounding and backing machine.
$_8D^c$	Await rounding and backing.
P_8^c	Round and back 50,000 books. Stack on skids.
D_8^c	Await move to casing-in.
M_9^c	Move skids to casing-in machine.
$_9D^c$	Await casing-in.
P_9^c	Case-in 50,000 books. Deliver to building-in.
P_{10}^c	Stack cased-in books on building-in boards. Compress each stack.

D_{10}^c Wait 24 hours for adhesive to set.

M_{11}^c Move to inspection station.

$_{11}D^c$ Await inspection.

P_{11}^c Inspect 50,000 books and apply jackets by hand. Stack on skids for mailing.

D_{11}^c Await move to mailing.

M_{12}^c Move to mailing station.

$_{12}D^c$ Await mailing.

P_{12}^c Insert 50,000 books into formed and addressed cartons; machine seal; place on conveyor to post office.

NOTE: The superscript c denotes conventional technology.

Table 7.2 Abridged flow process chart: Articulated technology

P_1^a Print 250,000 copies of ten 32-page forms (two up). Slit, slice, and fold 500,000 sheets into 32-page signatures. Stack on skids.

D_1^a Await move to gathering machine.

M_2^a Move skids of signatures to gathering machine. Also move skids of 50,000 front and 50,000 back endpapers to gathering machine.

$_2D^a$ Await gathering.

P_2^a Gather 50,000 books of ten 32-page signatures plus two 4-page endpapers. Clamp, slice off bind edge, rough, glue, dry in binding machine. Trim three sides and stain top edge of each book. Stack on skids.

D_2^a Await move to rounding and backing machine.

M_3^a Move skids to rounding and backing.

$_3D^a$ Await rounding and backing.

P_3^a Round and back 50,000 books. Stack on skids.

D_3^a Await move to casing-in.

M_4^a Move skids to casing-in machine.

$_4D^a$ Await casing-in.

P_4^a Case-in and build-in 50,000 books. Stack on skids.

D_4^a Await move to jacketing.

M_5^a Move to jacketing machine.

$_5D^a$ Await jacketing.

P_5^a Jacket 50,000 books. Inspect samples. Stack on skids.

D_5^a Await move to mailing.

M_6^a Move to mailing station.

$_6D^a$ Await mailing.

P_6^a Insert 50,000 books into formed and addressed cartons; machine seal; place on conveyor to post office.

NOTE: The superscript a denotes articulated technology.

for affixing adhesive along the bind edge of $2 \times 50{,}000 = 100{,}000$ endpapers.

As a final assumption, both flow process charts begin with press-work and end with mailing. Since preceding and succeeding oper-ations could be alike in both technologies, these abridged charts are intended to show the essential differences between the two systems.

Before comparing Tables 7.1 and 7.2 in detail, it is important to remember that what has been called the "conventional" technology is, although mechanized, essentially a jobbing system, one that is capable of printing books of diverse kinds and sizes, on different kinds and weights of paper, in different inks, with signatures gath-ered and bound in different ways. The articulated system has sac-rificed this capability in order to do but one thing: make books that are, in essence, alike.

It requires only a glance at Tables 7.1 and 7.2 to perceive that the sacrifice of variety for sameness has yielded very large gains. The total time (equation [1.1]) required to produce the assumed job by the conventional technology is the sum of 10 moves, 20 delay intervals after and before processing, and 12 processes:

$$T^c = \Sigma \left(M^c_{2-9,11,12} + D^c_{1-8,10,11} + P^c_{1-12} + {}_{2-9,11,12}D^c \right).$$

In contrast, the total time required for the articulated system shows only 5 moves, 10 delays, and 6 processes:

$$T^a = \Sigma \left(M^a_{2-6} + D^a_{1-5} + P^a_{1-6} + {}_{2-6}D^a \right).$$

Whatever difference there may be between T^c and T^a depends, of course, upon the magnitudes of the components as well as the number of them. Moving times are not likely to be of much signif-icance, but delays, as has been shown for jobbing systems generally, are certain to be much greater for the conventional method. Not only are there 20 such intervals for D^c, compared to only 10 for D^a, but the approximate balance that goes with articulation would be likely to make each elemental time less than the elements that compose total delay time D^c.

Processing intervals, most costly of the four time elements, also favor articulation. Invention of the flexible adhesive permitted elim-ination of the tedious and expensive task of sewing; this is a change in process rather than in system technology, but the innovation has permitted the kind of intercenter balance that has made articula-tion—and some attendant automated and continuous processing—possible.

Additionally, as suggested previously, sameness in the articu-lated method has reduced, and to a considerable degree eliminated, changeover requirements between one book and another. Some interform and interjob adjustments remain, but they are much less time-consuming than the paper and ink changes and size adjustments that must be made throughout the jobbing system.

Thus, we can say that P^c is greater than P^a. From these speculations we can also say that T^c is much greater than T^a: $T^c \gg T^a$.

This conclusion about the advantages of T^a over T^c applies also to cycle time (equation [1.2]), working-capital requirements, in-process inventory, and floor-space requirements (equations [1.5], [1.6], and [1.7]), all of which are much reduced in the articulated technology. We have no suitable data for the numerators of the equation for system efficiency (equation [1.4]), but since the denominators involve the sums already described, it is clear that system efficiency has been remarkably improved by articulation.

Against these considerable advantages of articulation over the jobbing tradition, fixed-capital costs and the cost of stoppages must be weighed; so also must the probable longevity of the market for books of the kind described.

> The fixed-capital needs probably were modest. Linked presses and folding machines are fairly common, and gathering, rounding and backing, and casing-in machines are standard. The slicing, roughing, gluing, drying, trimming, and staining sequence evolved from the in-house adaptation of a standard binding machine, with articulation of component steps.

> There being such small inventories of work in process at so few separate work centers, stoppages at any one center could be expected to bring the system to a halt. No information about the frequency or duration of holding time (H) is available in this case, but it seems likely that such interruptions were tolerable.

> The emergence and growth of paperback demand may well have changed the market. If so, it seems probable that adaptation of articulated technology has been practicable.

This example of articulation bespeaks imagination and ingenuity in book manufacturing, an industry with a jobbing tradition as old as the books produced using the movable type of Johann Gutenberg.

EIGHT

Balanced and Continuous Systems

Provision of a single good or service has been shown to make possible the sequencing, juxtaposing, and approximate balancing that have characterized articulation. In the language of evolution, the mutation from jobbing to articulated systems has resulted in the creation of new species, dependent upon single sources of sustenance, and prospering as long as these remain sufficient.

As shown by the assumptions of Table 6.1, imbalances require flexibility in starting and stopping component work stations and, even under assumed conditions of operational "perfection," lead to in-process delays and sequencing idleness. These intervals are small compared to those attending jobbing systems, but they are not inconsiderable. Through juxtaposition of work centers and work stations, moving times also are small, but they too are positive.

Mutations from the kind of articulation described have not been so dramatic as the transition from jobbing to articulation, but evolutionary changes have taken place: *balanced systems* have resolved problems of in-process delays and sequencing idleness, and *continuous systems* have demonstrated that it is sometimes possible to process goods in transit from one work station to another.

Assumptions previously made to characterize the progression of Job. No. 1 through the single-channel, five-stage system will be modified to show the effects of balancing and continuity.

Balanced Systems

Table 2.2 is a flow process chart for Job No. 1, with data assumed to typify jobbing system technology. Table 6.2 shows intervals assumed to represent progression through a system articulated for the processing of that single good or service.

To compare these hypothetical sequences with a balanced technology, imagine that Job No. 1 is divisible into components according to the intervals shown in Table 8.1. All but one of the processing intervals assumed in this table are smaller than the values of P^1 previously assigned to each of the five work centers.

By way of explanation, this sequence assumes that the operator at Work Center 1 carries each unit from the point of entry to his station and there, without preoperation delay ($_1D^1 = 0$), processes it in the same 4-minute interval assumed for P_1^1 in Table 6.2. He then, again without delay ($D_1^1 = 0$), places the unit on some form of conveyor, which, in an interval assumed to require 0.5 minute, moves the work to the successor station, designated by the subscript 21. One unit will reach this work station from Work Center 1 every 5 minutes and there, assuming perfect balance, will be processed in exactly that same time ($P_{21}^1 = 5$ minutes).

This same routine repeats throughout the table, with half-minute moving intervals between 1 and 21, and between 21, 22, 23,

Table 8.1 Moving and processing intervals

$M_1 = 0^a$	$M_{23} = 0.5$	$M_{32} = 0.5$	$M_{43} = 0.5$	$M_{46} = 0.5$	$M_{53} = 0.5$
$P_1 = 5$	$P_{23} = 5$	$P_{32} = 5$	$P_{43} = 5$	$P_{46} = 5$	$P_{53} = 5$
$M_{21} = 0.5$	$M_{24} = 0.5$	$M_{41} = 0.5$	$M_{44} = 0.5$	$M_{51} = 0.5$	$M_6 = 0.5$
$P_{21} = 5$	$P_{24} = 5$	$P_{41} = 5$	$P_{44} = 5$	$P_{51} = 5$	
$M_{22} = 0.5$	$M_{31} = 0.5$	$M_{42} = 0.5$	$M_{45} = 0.5$	$M_{52} = 0.5$	
$P_{22} = 5$	$P_{31} = 5$	$P_{42} = 5$	$P_{45} = 5$	$P_{52} = 5$	

Summary

$$\sum_i P_j = P_1 + P_{21-24} + P_{31,32} + P_{41-46} + P_{51-53} = 5 + 20 + 10 + 30 + 15$$
$$= 80 \text{ min.}$$

$$\sum_i M_j = M_1 + M_{21-24} + M_{31,32} + M_{41-46} + M_{51-53} + M_6 = 0 + 2 + 1 + 3$$
$$+ 1.5 + 0.5$$
$$= 8 \text{ min.}$$

NOTE: Moving and processing intervals are assumed to be in balanced sequence, Intervals for delays and idleness are assumed to be zero and therefore are not included. Work Stations 21–24 correspond to Work Center 2 in Table 6.1; Work Stations 31 and 32 to Work Center 3; 41–46 to Work Center 4; and 51–53 to Work Center 5. Superscripts denoting job number have been omitted, since all units are the same. All times are in minutes.

[a]Move M_1 is made by the operator and therefore is included in P_1.

and 24, 31 and 32, 41, 42, 43, 44, 45, and 46, and 51, 52, and 53, and exit from the system. Five-minute intervals have been assumed as the time for P^1 at each station. There are no delays or periods of idleness throughout the perfectly balanced system.

SYSTEM CHARACTERISTICS

Before comparing the performance of this sequence with data assumed for jobbing and articulation, let us consider certain characteristics of balanced system technologies.

Balance between work stations sufficiently approaching the "perfection" that has been assumed here is very difficult to achieve and, once achieved, is comparably difficult to maintain. The applicability of balanced system technology, depending as it does upon constancy of human performance, is therefore much more limited than articulation, and is restricted most often to assembly operations.

Operational continuity, shown to be important in articulated systems, is of even greater significance in balanced systems. In articulated systems, as shown in Table 6.1, there are at times some in-process inventories, but in balanced systems there are none: when work at one station halts, so does the entire sequence.

To prevent balanced system idleness, caused by stoppages or performance variances, "roving operators," trained to perform a variety of tasks, are sometimes employed to prevent or relieve backlogs. Such operators can enhance operational continuity but at increased cost.

In contrast to overstaffing and overequipping in jobbing and articulated systems in order to compensate for variations in demand, imbalances, and breakdowns, the provision of roving operators or backup equipment in balanced systems can be characterized as planned redundancy. Perhaps this marked the beginning of conscious cognizance of redundancy as a concept of great importance in the design of many systems, particularly those in which continuity or safety are paramount—for example, electricity generation, air and space flight, and the like.

Balanced systems can utilize the concept of "machine pacing," whereby the speed of a moving belt or conveyor can be
regulated to ensure rapid and diligent operator performance.
Use of this technique, plus proficiency born of repetition,
usually results in lower values for P in balanced systems.
However, because of allegations of enforced "speed-ups,"
machine pacing has been the cause of much controversy.

Control of balanced systems is similar in kind but relatively
better in degree than control of articulated technologies.
Once started at the first work station, work in a balanced
system can only proceed to successive stations and emerge
finally from the last.

The fixed capital required to design, equip, train, and test
balanced systems is relatively greater than that required for
articulated and jobbing technologies.

These characteristics portray balanced system technology as a
refinement of articulation, a mutation of lesser magnitude than the
change from jobbing to articulation. But even though limited in
applicability, balancing has valuable uses, as the following comparisons show.

SYSTEM COMPARISONS

Total time and cycle time

Since Table 8.1 shows a single-channel sequence, total time T^1 and
cycle time C^{1*} (equations [1.1] and [1.2]) once again are alike:

$$T^1 = C^{1*} = 8 + 0 + 80 + 0 = 88 \text{ minutes.}$$

This number is less than the 162-minute interval for articulation
and very much less than the 916 minutes required for processing in
the jobbing mode. Because all the balanced work stations are in
series, there are more moving intervals, but each has been assigned
only 0.5 minute by the conveyor assumption. All delays (66 minutes
in the articulated case) have been eliminated, and processing time
has been reduced by assuming that the pace of balanced productivity
will be increased.

Operational efficiency

In order to show the effects of adding roving operators to the system, operational efficiency will be calculated on a daily basis rather than for a single unit.

The assumptions of Table 8.1 show an output rate of one unit each 5 minutes, or 96 units per 8-hour day (480/5 = 96). Based upon this daily output rate and again using 94 standard minutes as the output value of one unit of product or service, operational efficiency for the complement of 16 operators shown in equation (1.3) is:

$$(E_p)_{480}^{**} = \frac{94 \times 96}{(96 \times 8) + 0 + (16 \times 480)}$$

$$= 9{,}024/(768 + 7{,}680) = 9{,}024/8{,}448$$

$$= 1.068 = 106.8\%.$$

(This ratio yields the same quotient as the operational efficiency for a single unit: $E_p^1 = 94/(8 + 0 + 80) = 1.068.$)

The addition of one roving operator to prevent system idleness would add 480 minutes to the denominator of equation (1.3) and reduce operational efficiency to 9,024/(8,448 + 480) = 9,024/8,928 = 1.011. A second roving operator would further reduce operational efficiency to 9,024/9,408 = 0.959. For either one or two roving operators it has been assumed that idle time due to stoppages would be kept at zero, a perfection seldom, if ever, realized in balanced systems.

System efficiency

Again using Δt = 480 minutes and 16 operators, system efficiency (equation [1.4]), because intervals for D and I are assumed to be zero, is the same as the operational ratio calculated above:

$$(E_s)_{480}^{**} = \frac{94 \times 96}{(96 \times 8) + 0 + 0 + (16 \times 480) + 0}$$

$$= 9{,}024/8{,}448 = 1.068 = 106.8\%.$$

Working-capital requirements, in-process inventory, and floor-space requirements

Since there are neither pre- nor postoperation delays nor any idleness due to sequencing, and since floor-space requirements do not vary, the intervals proportional to working-capital requirements, in-process inventory, and floor-space requirements are the same and can be represented by the same sum of M_j^1 and P_j^1 shown above for total time and cycle time:

$$W^1 \sim V^1 \sim A^1 \sim 8 + 80 = 88 \text{ minutes.}$$

As in the case of jobbing and articulated comparisons, these calculations are based upon oversimplified assumptions, but the magnitudes of intersystem differences are sufficient to conclude that balanced systems, when feasible, have valuable attributes.

Continuous Systems

A continuous system, as said before, is one in which processing is carried on while products are moving through the system, a denotation that in effect combines the elemental times for moving and processing into single processing intervals. Continuous processing usually reduces the duration of the combined intervals, but this is not a necessary condition.

What is necessary is that there be a stream of product that can be worked upon (heated, cooled, mixed, filtered, precipitated, drawn, reacted) while moving. Chemical and metallurgical processes, as in the cracking of petroleum, distillation of spirits, continuous casting or rolling of metals, or drawing of wire, have been most suitable for continuous system technology, while service systems have not. Sometimes, however, discrete processing events take place so rapidly and repetitiously as to partake of the character of a continuous stream, and interunit balances approach continuity.

SYSTEM CHARACTERISTICS

Balanced systems have been described as mutations of articulation, and, at least in fancy, continuous systems that must be balanced as well can be viewed as mutations from balanced system forebears. Be that as it may, continuous systems bring new attributes to the system characteristics previously stated:

> For both fixed and operating capital, systems of all kinds trend away from labor intensity toward capital intensity. Allowing for the continuing trend of substituting capital for

labor, jobbing systems, articulated systems, and balanced systems operationally can still be either labor or capital intensive but it is difficult to imagine a continuous system in which operations are *not* capital intensive. Given the requirement that there be a stream of product or its equivalent, continuous systems must contrive "processing in motion," a characteristic that is likely to exceed human capabilities. Human input to continuous systems must therefore be transferred to machines or, more properly, instruments to regulate the processing that is going on during passage of the product through the system. In continuous systems the human contribution has not only been reduced absolutely but changed relatively: from manual performance to the monitoring of instruments.

Control of processing in motion by instruments in order to achieve continuity has accelerated understanding and utilization not only of redundancy but also of control theory and the principle of feedback. A human monitor, provided with measurements showing the state of the system, can exercise judgment as to when the system is off course and take corrective action, but it is much better to establish control limits, the exceeding of which, by feedback, will provide for instrumental correction. Both of these developments, fostered by the evolution of continuous systems, have paved the way for automation, which will be considered in the next chapter.

Shifts from performing to monitoring have brought about profound changes in the personnel requirements of continuous systems. Activity has given way to inactivity, and manual skills and dexterity have been subordinated to the need for technological understanding and the ability to function under stress, so that out-of-control situations can be dealt with promptly and correctly. The repetitious monotony of the "man on the assembly line" has become a different but equally insidious kind of monotony: inactivity or contrived activities having the flavor of "make work," punctuated only rarely by out-of-control incidents and, even more rarely, by crises, the consequences of which can be dire. The control room of a nuclear power plant provides an example of the need for unremitting alertness for untoward, rarely occurring events, followed at once by just-right corrective measures.

Conversion to continuous technology frequently involves changes from processing in batches or units to processing continuously. Such changes often require the development of new processing technology and instrumentation for process control.

SYSTEM COMPARISONS

Time intervals assumed for the processing of Job No. 1 through a jobbing system have been changed, first, to intervals for an articulated system and then to those for a balanced technology. A degree of verisimilitude has been claimed for the initial assumptions and for the changes from them.

Because a continuous system, existing in "pure" form, becomes a single processing unit, values of M, D, and I become zero, leaving only P to represent the interval between raw-material input and finished-product output. Output, even though emerging in a continuous stream, must be measured in discrete units—for example, barrels of petroleum products, gallons of whiskey, feet of wire.

For P to represent Job No. 1, we must assume a continuous process interval for P^1, an assumption that could be anybody's guess. Perhaps as good an assumption as can be made to preserve a modicum of verisimilitude is to say that all the moves (8 minutes) assumed for the balanced sequence are embedded in the processing intervals without any change in them. Assume, then, that P^1 for the continuous system is 80 minutes. Since M^1, $_1D^1$, D_1^1, and I^1 are all zero, this means that total time T^1, cycle time C^{1*}, and the proportionalities for working capital, W^1, in-process inventory, V^1, and floor space, A^1, all have this same value.

If we also continue to assign 94 standard minutes as the output value of a unit of product, both operational and system efficiency will then be measured by the ratio $94/80 = 1.175 = 117.5$ percent.

Having said that continuous systems are capital intensive, and having ignored the economies of scale that usually attend the creation of continuous systems, the choice of $P^1 = 80$ is probably much too conservative. However, the purpose here is to make comparisons among system technologies, and for that these assumptions may serve.

Summary comparisons

In this chapter and in preceding chapters on jobbing and articulated systems, intervals for what has been called Job No. 1 have been assumed to show system performance according to the characteristics described for each of four types of operational systems: jobbing, articulated, balanced, and continuous.

For the first of these, jobbing system capability for a *variety* of goods and services was a principal attribute, but for each successor system, variety was assumed to give way to *sameness*: articulated, balanced, and continuous systems were assumed to produce *only* Job No. 1. The trade-off for improved system performance therefore involved a sacrifice of system versatility.

In later chapters we will see this trend toward product and service concentration begin to be reversed by system technologies that regain degrees of versatility without sacrificing performance, at the same time avoiding the high costs and severe system disruptions that attend product changes in articulated, balanced, and continuous systems.

Unfortunately, values accruing to versatility—the capability to do different things—affect not only the disutilities attending changeovers but also the attractiveness of versatility to customers and clients, and the magnitudes and longevities of markets. None of these—changeover costs, attractiveness to customers and clients, or market extensions—can be measured and compared by using elemental times for M, D, P, and I, as they have hitherto been assumed for fictitious Job No. 1.

Succeeding chapters will not attempt to quantify the benefits of versatility to customers, clients, and managers, but will show that the disutilities that once attended diversity—the high costs and disruptions of changeovers—can be greatly diminished by controls that use physical, mathematical, or programmatic models.

However, before abandoning Job No. 1, let us summarize the comparisons for jobbing, articulated, balanced, and continuous technologies given in Table 8.2.

The intervals compared in Table 8.2 show that each successor system has operational advantages over predecessor technologies. However, these advantages in speed, efficiency, and economy are mitigated by restrictions that must be satisfied if a net gain is to be realized:

Table 8.2 Jobbing, articulated, balanced, and continuous systems: A comparison

Elements, Summations, Ratios			Jobbing	Articulated	Balanced	Continuous
$\sum\limits^{j} M_j^1$		=	36	8	8	0
$\sum\limits^{j} {}_j D^1$		=	686	66	0	0
$\sum\limits^{j} I_j^1$		=	0	3	0	0
$\sum\limits^{j} P_j^1$		=	94	88	80	80
$\sum\limits^{j} D_j^1$		=	100	0	0	0
Total time	T_1	=	916	162	88	80
Cycle time	C^{1*}	=	916	162	88	80
Operational efficiency	E_p^1	=	72%	95%	107%	118%
with 1 roving operator					101%	a
with 2 roving					96%	a
operators						
System efficiency	E_s^1	=	10%	57%	107%	118%
with 1 roving operator					101%	a
with 2 roving						
operators					96%	a
Working capital						
requirements	W^1	~	916	165	88	80
In-process inventory	V^1	~	916	162	88	80
Floor space requirements	A^1	~	916	570	88	80

NOTE: Comparisons are based upon assumed intervals for processing one unit of Job No. 1.

[a]Redundant equipment needed to ensure operational continuity would require additional capital and incur additional cost. This would, in effect, reduce operational and system efficiency. No attempt has been made here to estimate cost of this kind.

Articulated operations technology *is* better than jobbing, but *only* if product or service concentration is feasible, if market prospects are sufficiently promising and enduring, and if the capital and transitional costs are warranted by the advantages of a fixed sequence of juxtaposed work centers in approximate balance.

Analogously, balanced operations technology *is* better than articulation *only* if the capital and transitional costs of attaining and sustaining operational and logistical balance, avoiding stoppages, and ensuring a stable market are less than the expected operational gain.

The refinement of continuous operations technology *can* yield net gain, but the restrictions are even more severe than those for balanced systems: processing in motion may not be possible, and even if it is feasible, substantial capital investment may require a large market, one that is sufficiently stable to prevent or postpone costly changeovers.

The intervals in Table 8.2 therefore do not permit a conclusion of technological superiority for any of the four systems compared *except* in the context of the particular purposes for which the individual system exists. There is, however, potential value in understanding the characteristics of each operations technology.

Most systems—nearly all in fact—are "hybrids." It is hard, perhaps impossible, to imagine a "pure" continuous system, one that does not have jobbing appendages for maintaining the system's continuity. Conversely, there are likely to be subsystems within an otherwise "pure" jobbing network that can be articulated or perhaps even balanced to good effect, with ancillary benefits deriving from improved controls and diminished division of labor. Improvements of these kinds can be realized by periodic—and to an extent continuous—analysis of the time intervals for moving, delays, processing, and idleness that comprise operations technologies.

NINE

Batch to Continuous System Conversions

GLACÉ FRUIT
TELEPHONE CABLE

The bookmaking system described in Chapter 7 combined articulation, balance, and automation, with but one sequence involving a continuous process. All of these were evident in the gather-slice-roughen-glue-dry-trim-stain sequence, but emphasis was upon articulation, and only incidentally upon continuity.

The examples that follow also involve hybridization, but the emphasis is different: it is upon conversions, with beneficial results, from batch processing to continuous sequences. The examples involve manufacturing (it is hard to imagine a continuous process that does not), but the products, food and cable, are quite different.

GLACÉ FRUIT

Glacé fruit[1] is used in various ways in households and bakeries, most familiarly as an ingredient in fruit cake. The raw material from which glacé fruit is processed is harvested before the fruit is ripe, and the not-quite-ripe crop is then put into wooden barrels, immersed in sulfur brine to bleach and preserve the fruit for later, seasonal use.

During storage, the fruit is permeated with sulfur, which must be removed and, in effect, replaced by sugar. This is done by repeatedly washing the fruit with water, draining away the water, and replacing the water with syrup of increasing sugar content.

---

1. Batch and continuous technologies for the processing of glacé fruit were used, in the order described here, by C. M. Pitt & Sons Company in Baltimore, Maryland. I am indebted to Mr. W. S. Arnold for details of both system technologies (see note 4).

Batch processing

Equipment for the batch process consisted chiefly of a shallow rectangular tank to contain the fruit in process; a kettle to contain the mixture of sugar, corn syrup, and water; a vacuum kettle for the evaporation of excess water; and a reversible transfer pump, connected as shown schematically in Fig. 9.1.

The sequence of processing steps began with placement of approximately 1,000 pounds of fruit into the tank from barrels brought from storage.

The removal of sulfur was accomplished by washing the fruit with water, draining away the sulfur-laden water,[2] and then repeating the washing-draining cycle until the sulfur content of the fruit became sufficiently low.

Concurrently with these cleansing cycles, a mixture of corn syrup, sugar, and water, having an overall sugar content of about 15 percent, was prepared in the kettle for use when the removal of sulfur had been accomplished.

The second exchange process, this time involving sugar for water, was then begun by putting syrup from the kettle into the tank of now-water-laden fruit. As the sugar-for-water exchange took place, it became necessary to remove excess water and to increase the sugar content of the syrup. This was done by pumping liquid from the tank into the vacuum kettle to evaporate excess water,[3] and by adding corn syrup and sugar to increase the sugar content of the mix, which was then returned to the tank.

As before, these exchange cycles were repeated until the sugar content of the fruit had reached the desired high level. Each cleansing and sweetening cycle required about 24 hours, and the entire procedure for each batch lasted about 21 days.

Each completed batch was dipped from the tank onto screens and allowed to stand until sufficiently dry for packaging.

To prevent loss of fragile product, careful controls and frequent monitoring were necessary.

Washing with water tends to make fruit soft, such that layers at or toward the bottom of a deep tank could be mashed by the weight of the fruit above. Thus, a shallow tank was used so that the

---

2. At the time these systems were employed, cleansing water from the tank was drained into the sewer. Since the water contained sulfuric acid, this mode of disposition would now be forbidden, unless the acid in the waste water was first neutralized.

3. Low-pressure evaporation, rather than evaporation by heat, was the method of choice, because an elevated temperature would crystallize the sugar.

Fig. 9.1 Schematic diagram of equipment used for the batch processing of glacé fruit.

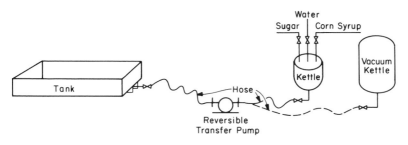

column of fruit in each batch would be relatively wide and long but not high.

Sulfur content of the fruit was measured after each of the several cleansing baths to determine when washing with water should give way to the application of syrup.

During subsequent sequences of applying syrup, as sugar replaced water in the fruit, sugar content also was monitored, to determine when and how often syrup should be returned to the vacuum kettle for removal of excess water, followed by enrichment of the syrup and its return to the fruit in the tank.

Each time the tank was drained, either for replacement of fresh water or enriched syrup (more than 20 times for each 1,000-pound batch of fruit), crushing of soft fruit at the orifice of the tank caused some spoilage.

Using the notation specified in Chapter 1 and the superscript b to denote the batch process, the following preparatory events were performed only once for each batch:

M_1^b Move barrels of fruit to tank.

P_1^b Open barrels and put 1,000 pounds of fruit into tank.

P_2^b Prepare mixture of sugar, corn syrup, and water to a 15 percent sugar content.

These initial operations were followed by cycles of the steps needed to remove the sulfur from the fruit:

M_3^b Add cleansing water to tank.

D_3^b Allow to stand.

M_4^b Drain and dispose of cleansing water.

P_4^b Measure sugar content of fruit.

Repeat cycle.

After the sulfur was removed from the fruit, the following cycles were repeated:

M_5^b Add syrup to fruit in tank.
D_5^b Allow to stand.
P_5^b Measure sugar content of syrup.
M_6^b Return syrup diluted with water to vacuum kettle.
P_6^b Evaporate excess water.
P_7^b Add corn syrup and sugar to syrup to increase its sugar concentration.

Repeat cycle by returning enriched syrup to the tank, allowing it to stand, and so on.

Final operations, performed once for each batch, were:

P_8^b Dip fruit from tank onto screens.
D_8^b Allow to stand.
P_9^b Package.

Using these elements and assuming that there were three cleansing cycles and 17 applications of syrup, the approximate total time T (equation [1.1]) for each batch of fruit can be represented by

$$T^b = M_1^b + P_1^b + P_2^b + 3(M_3^b + D_3^b + M_4^b + P_4^b)$$

$$+ 17(M_5^b + D_5^b + P_5^b + M_6^b + P_6^b + P_7^b) + P_8^b + D_8^b + P_9^b.$$

Cycle time C^{b*} (equation [1.2]) would be slightly less than the time required for T^b, by the interval P_2^b, performed concurrently during the cleansing cycles. The above expression for T overstates total time by an unknown but not large amount, because not all the enriching cycles involved the evaporation of excess water.

Continuous processing

Improvement of the method just described did not change the processing of the fruit in 1,000-pound batches, nor did it alter the replacement of sulfur by water and the water by sugar. Improvement did, however, make continuous the successive applications of cleansing water and, by use of heat in the tank, speed the removal of sulfur from the product, at the same time contriving the continuous preparation of incrementally sugar-rich syrup in the tank. The result was a reduction in start-to-finish time for each batch from 21 days to about 5.

Fig. 9.2   Schematic diagram of equipment used for the continuous processing of glacé fruit.

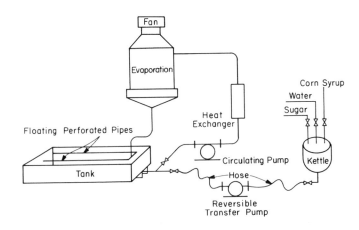

To speed the removal of sulfur from the brine-preserved raw material, heating coils were placed in the tank, not only to warm the cleansing water and enhance the dissolving of the sulfur but also to gasify and vent the decrement of sulfur oxide (SO) and sulfur dioxide (SO_2).

In the earlier system, vacuum had been used to evaporate excess water when syrup was returned to the vacuum kettle to bring the syrup to the desired higher sugar concentration. In the new method, kettle cycles were made continuous, as suggested in Fig. 9.2.

Starting with the initial sugar concentration of 15 percent, syrup was applied to the fruit from above, after cleansing water had been drained away. From the bottom of the tank, as the sugar content of the syrup fell to a threshold level, the low-sugar syrup was pumped into and through a heat exchanger and thence to the evaporator, to discharge water vapor into the air and concentrate the syrup. As the remaining syrup flowed toward the bottom of the tank, the addition of sugar and corn syrup restored the sugar concentration to the desired higher level, and once again, after removal of the cleansing water, syrup was applied to the fruit in the tank.

In the earlier method, careful control of sugar and sulfur content of syrup and fruit were necessary to prevent loss of the perishable and fragile product. In the continuous method, more extensive and sensitive controls were necessary.

With the more rapid removal of sulfur by heating the fruit and cleansing water and by gasifying some of the sulfur, control of temperature and dwell time became more critical, necessitating the determination and establishment of permissible limits and instrumentation to bring changes when called for by control limits.

Analogous, more critical control limits and processing changes had to be established for flow rates, dwell times, water and sugar concentrations, and incremental enrichments of the syrup, these being augmented by monitoring of the fruit itself, to determine when sugar content had reached the desired terminal level before drying and packaging.

Comparisons

Because processing and moving are combined as products pass through "perfect" continuous systems, flow process charts, insofar as moves and delays are concerned, lose meaning. The system itself *is* a process.

To a less than perfect degree this is true in the case under consideration. However, the fact that batch time is 21 days in the earlier method and only 5 days for the continuous technology leads to reasonable conclusions.

Using the superscript c to denote the continuous method, there were:

Large reductions in total time T and cycle time C^{c*}:

$T^c \ll T^b$ and $C^{c*} \ll C^{b*}$ (eqs. [1.1] and [1.2]).

Large increases in operational efficiency E_p^c and system efficiency E_s^c:

$E_p^c \gg E_p^b$ and $E_s^c \gg E_s^b$ (eqs. [1.3] and [1.4]).

Large reductions in working-capital requirements W^c, in-process inventories V^c, and operational floor space A^c:

$W^c \ll W^b$, $V^c \ll V^b$, and $A^c \ll A^b$ (eqs. [1.5], [1.6], and [1.7]).

An additional advantage of the new method, not reflected in time measurements, derived from reduced spoilage from drainage losses in the original batch process.

In this case it seems apparent that fixed-capital requirements were modest, consisting chiefly of auxiliary equipment, transitional engineering costs, control instrumentation, and modifications to existing equipment.

Should the market justify a further step, it seems possible that the process could be made continuous throughout, from brine to finished product, by moving the fruit through some form of shallow channel in a system programmed to provide the necessary processes and controls.

TELEPHONE CABLE

In the manufacture of telephone cable[4] the basic elements were— and to a considerable extent still are—a pair of copper wires. In- sulation applied to each single wire was in one of ten different colors; these were twisted into 25 different pair combinations, no two of which were alike (e.g., blue-white, red-yellow, black-white, etc.). The rotary motion used to twist each pair of wires was different for each of the 25 color combinations—that is, the number of twists per foot were the same for, say, all red-black pairs but different for every one of the other 24 paired color combinations, and each of these was different from all the others. These twist differences min- imized "parallelism" and consequent "cross talk," a transmission problem caused by induction. The voice fluctuations in telephone conversations cause current fluctuations that can induce like fluc- tuations in parallel wires, thereby transmitting an unwanted, as well as a wanted, conversation.

Typically, the 25-pair combinations were then stranded together and bound with a pair of tapes in two of the ten colors. Units of 12- and 13-pair combinations also were made and bound in the same two-color binding as the "parent" 25-pair unit. Each 25-, 13-, and 12-pair unit described an approximate circular configuration as it passed through the stranding and binding process.

These typical 12-, 13-, and 25-pair units were chosen to permit the stranding of various larger cable sizes. At the same time, as will be shown, by suitable strand combinations a circular configuration would be maintained for binding and sheathing into finished cable.

Color coding enabled linemen, repairmen, and others to identify each pair of twisted wires by their 25 different color combinations and to identify the unit in which each pair was bound by the colors of the tapes binding each unit. The use of 25 binding-color pairs and 25 paired wire-color combinations provided 625 identifiable differ- ences.

---

4. For information about the batch and continuous technologies for making cable, I am greatly indebted to Mr. W. S. Arnold, retired Department Chief, Engi- neering, of the Manufacturing Division, and to Mr. Lorne R. Guild, retired Senior Staff Engineer, both of the Western Electric Company.

Thus, in connecting wires to install lines, or to repair lines that have been severed, operators need follow but one explicit instruction: to ensure or restore circuit integrity, always connect each wire to another of the *same color*, in a pair of the *same color combination*, from a unit with the *same color binding*. And, like the Munsell color code, combinations could be specified orally or in writing almost anywhere.[5]

Batch processing

In the earlier batch process, the first operation was to draw 3/8-inch copper rod down to 12-gauge wire,[6] taking this thinner wire up on a reel weighing about 300 pounds. As a second step, this wire was taken off the reel and passed through a die to draw the wire to one of several thinner gauges (22-gauge will be assumed here). This thinner and longer wire was again taken up on a reel.

To relieve stresses and hardness caused by drawing, reels of 22-gauge wire were then annealed in an oxygen-free, "inert" atmosphere to prevent oxidation. Plastic insulation in one of the ten basic colors was then applied to the annealed wire, and the insulated wire was again taken up on a reel and temporarily stored to await twisting into pairs. Each reel of insulated wire weighed about 60 pounds.

To make any one of the 25-pair color combinations, insulated wire reels in each of the two desired colors (e.g., red and green) were then taken from inventory to a twisting machine, there to be paired in a twist pattern unique to that color combination as described above. The twisted pair was then taken up on a reel to await stranding into units.

Reels of the desired color combinations were next loaded into a "strander," which formed and bound the pairs into "units" com-

5. Color coding was, of course, essential to circuit integrity, and circuit integrity was essential to system integrity, if not to corporate survival. Error-free color discrimination was therefore very important. It was necessary to select the ten basic colors with great care, to choose color combinations that would not be confused under many different, often unfavorable, field conditions, and to avoid color blindness on the part of employees.

6. Gauge measurements of wire are confusing, to say the least. Gauge numbers are inversely related to wire diameters, so that smaller numbers specify larger diameters and vice versa. In this case, 12-gauge specifies a diameter of 0.0808 inch and 22-gauge a diameter of 0.0253 inch. However, in telephony, diameter is of less significance than electrical resistance, which may vary in ohms per foot for wire of constant diameter. Control to maintain constant resistance was of greater importance in the continuous method.

Fig. 9.3   Cross-section (not to scale) of a 100-pair telephone cable composed
of a central 25-pair unit surrounded by three 12- and three 13-pair units.
Stranding and binding of the smaller units around the central core has caused
the smaller units to assume elliptical shapes that yield a more nearly circular
configuration. In addition, the linear and rotational movement of the units
describes a circle tangential to the maximum outside diameter.

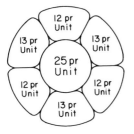

prising various numbers of pairs. These stranded units were then
taken up on a "core truck," a large vertical reel built on casters so
that it could be moved about. These combinations could comprise
from 6 to 100 pairs of wires, and each such stranded and bound unit
could be completed as the "core" of a cable or combined with other
stranded and bound units to form a larger cable.

   To combine units into cables of larger size, core trucks bearing
units of the correct size and color coding were moved to the "cabler,"
there taken off the core trucks, stranded and bound together, and
again taken up on core trucks to await transport to the next operation.

   As pointed out previously, all unit combinations stranded and
bound to form cable were required to have an approximately circular
configuration, as shown by the cross-section of Fig. 9.3, where a 100-
pair cable is composed of a central 25-pair unit surrounded by three
12- and three 13-pair units. While the cross-section does not describe
a perfect circle, the elliptical shapes assumed by the smaller cir-
cumscribing units as they are stranded and bound do approach a
circular configuration, and the twisting linear movement does describe a circle suitable for subsequent operations.

   Core trucks of stranded and bound units were then taken to the
"sheathing line," where cable was made by

A core wrap of polyester film tape
A binding of nylon thread
A corrugated aluminum tape, formed around the cable
A second binding of nylon

A black plastic jacket, applied by extrusion
Winding on a shipping reel

Reels of cable were then subjected to physical and electrical inspection and, if found satisfactory, were shipped or stored.

Using notation that may be familiar, this method is summarized in the following sequence, for later comparison with the successor technology. As before, the superscript b will be used to denote the batch process, and the superscript c, the continuous method.

M_1^b	Move 3/8-inch copper rod to be drawn.
$_1D^b$	Await drawing.
P_1^b	Draw 3/8-inch copper rod into 12-gauge wire. Take up on reel.
D_1^b	Await move to finer die.
M_2^b	Move to finer die.
$_2D^b$	Await drawing to finer gauge.
P_2^b	Draw 12-gauge wire to 22-gauge. Take up on reel.
D_2^b	Await move to annealing.
M_3^b	Move reel to annealing.
$_3D^b$	Await annealing.
P_3^b	Anneal in inert atmosphere. Take up on reel in annealer.
D_3^b	Await move to insulation.
M_4^b	Move to insulation.
$_4D^b$	Await insulation.
P_4^b	Insulate wire. Take up on reel.
D_4^b	Await move to twisting.
M_5^b	Move reels to twisting.
$_5D^b$	Await twisting.
P_5^b	Twist wires into pairs. Take up on reel.
D_5^b	Await move to strander.
M_6^b	Move reels to strander.
$_6D^b$	Await stranding.
P_6^b	Strand and bind pairs of wires into cable units. Take up on core truck.
D_6^b	Await move to cabler.
M_7^b	Move core trucks to cabler.
$_7D^b$	Await stranding and binding in cabler.

$P_7^\natural$ Strand and bind units in cabler.
    Take up on core truck.
$D_7^\natural$ Await move to sheathing.
$M_8^\natural$ Move core trucks to sheathing.
$_8D^b$ Await sheathing.
$P_8^\natural$ Core wrap, bind with nylon thread, form aluminum tape, bind again with nylon, apply plastic jacket, and wind on shipping reel.
$D_8^\natural$ Await move to inspection.
$M_9^\natural$ Move shipping reel to inspection.
$_9D^b$ Await inspection.
$P_9^\natural$ Inspect electrically and physically.
$D_9^\natural$ Await move to shipping or stores.
M_{10}^b Move to shipping or stores.

Engaged in the single purpose of making telephone cable, this system was for the most part articulated and in approximate balance. However, because of color, twisting, and stranding differences, there were occasional inventory and operational complications, with delays like those characteristic of jobbing systems.

Continuous processing

The first step in the continuous process, as before, was to draw 3/8-inch copper rod down to 12-gauge, but for this method the wire was reeled on a much larger "stempack," one capable of holding about 3,000 pounds.

From the stempack the wire was unwound at the insulator, where it was first drawn to the desired thinner guage (e.g., 22-gauge), and next passed—continuously—into a strand annealer that heated the wire by DC current in the presence of steam to prevent oxidation. The moving wire was then reheated to the proper temperature for bonding the plastic insulation, and the plastic was also kept under temperature control. The annealed, insulated, and bonded wire was taken up on a reel.

The plastic was clear, not colored as in the batch process. Color in the form of pigment was added at the extruder inlet and mixed during the extrusion process, so that any one of the 25 basic insulation colors could be applied continuously and, as necessary, changed from one color to another "on the fly," with minimum scrap.

The insulated and colored wire was then passed through water moving in a trough, to cool and set the plastic, and the cooled wire was then reeled. During this annealing and cooling process, control was devised to maintain constant electrical resistance. Capacitance was controlled by use of a water-tube monitor during passage of the wire through the quench trough. To control the quench point of the hot plastic, signals from the monitor actuated a "stepping motor," a servomechanism that moved the end of the cooling trough toward or away from the output end of the plastic extruder. There was also a spark tester to measure and control insulation resistance.

Reels of wire, insulated in each of the ten basic colors, then proceeded to the twisting, binding, and stranding stations as before, this time in another continuous sequence.

To enter this sequence, wires are not taken from reels by rotation of the reels themselves. Instead, wires "fly off" as indicated in Fig. 9.4, a procedure that permits continuity by joining the end of the wire from Reel 1 to the beginning of the wire from Reel 2, and so on.

To make 100-pair cable, stranded and bound in the configuration shown in Fig. 9.3, 200 wires in the 25 color combinations previously described were twisted into pairs and these differently twisted pairs were then passed through seven "face plates" of the kind suggested in Fig. 9.5. These were arranged in a configuration like that shown in Fig. 9.3. In unison, these face plates rotated in alternately clockwise and counterclockwise directions to impart to each combination its own "SZ" twist to avoid parallelism. Each of the seven resulting 25-, 12-, and 13-pair units were bound in suitable colors and these bound units were passed through another, larger face plate, which also rotated in alternating clockwise and counterclockwise direc-

Fig. 9.4   Schematic diagram of stempacks, wire fly-offs, and an interstempack splice.

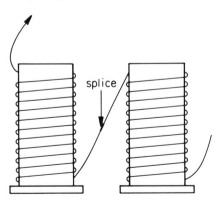

Fig. 9.5   Schematic arrangement (not to scale) of a typical face plate for stranding and binding 12-, 13-, and 25-pair units. In addition to the single hole in the center, there are 12 holes in the outer circle (circle A), 9 in the adjacent circle (circle B), and 3 in the inner circle (circle C). To make 25-pair units, all 25 holes are used; to make 12-pair units, all holes in circles B and C are used; to make 13-pair units, all holes in circle C and 10 holes in circle A are used.

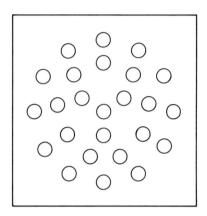

tions (Fig. 9.6).[7] The resulting 100-pair assemblage was then bound and taken up on a stempack.

All of these operations were performed continuously.

The stempack was then taken to sheathing, where a black plastic jacket was applied and the finished cable was wound on shipping reels, subjected to electrical and physical inspection, and stored or shipped. For cable of sizes larger then 100 pairs, additional stempacks of units could be used and formed into cable at the supply end of the sheathing line.

Elements of the flow process chart used for the batch method represented 300 pounds of wire drawn to 22-gauge and reeled. The continuous process began with a much larger batch: 3,000 pounds drawn first to 12-gauge, reeled on a stempack, and then continuously drawn to 22-gauge, annealed, insulated, cooled, and taken up on a reel. Bearing this yield difference in mind, the continuous sequence may be described as follows:

7. Face plates may be likened to the physical analogs that tools follow in metal working: they are physical "programs" that select desired pair combinations, choose and bind combinations together, and give a circular configuration to the units to be sheathed into cable. As in the case of other analogs, process changes can be made by changing face plate arrangements.

Fig. 9.6 Schematic representation of a "gathering" face plate in which seven units are twisted and bound into 100-pair cable. The number of pairs assigned to each hole is indicated by the number assigned to each circle. Like Fig. 9.5, this drawing is not to scale; the holes and the face plate itself have been made larger simply to suggest that the seven larger units require larger holes and a larger face plate than those used for pair combinations. In actual practice, in order to handle other cable combinations, face plates contain more holes than are shown.

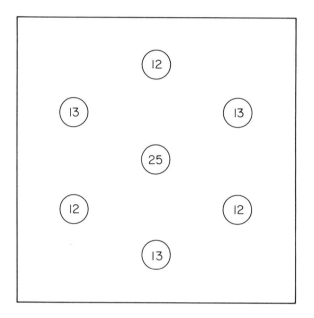

M_1^c Move copper rod to be drawn.

$_1D^c$ Await drawing.

P_1^c Draw 3/8-inch rod into 12-gauge wire.
Take up on stempack.

D_1^c Await move to insulator.

M_2^c Move stempack to insulator.

$_2D^c$ Await insulation.

P_2^c Draw to 22-gauge, anneal, reheat wire, heat plastic, apply plastic foam, cool in quench trough.
Take up on a reel.

D_2^c Await move to twisting, binding, and stranding.

M_3^c Move reels to twisting, binding, and stranding.

$_3D^c$ Await twisting, binding, and stranding.

P_3^c Fly wires off reels, twist into pair colors, pass through face plates, strand and bind into 12-, 13-, and 25-pair units, pass units through second face plate to form 100-pair cable, wrap and bind.

Take up on stempack.

D_3^c Await move to sheathing.

M_4^c Move stempack to sheathing.

$_4D^c$ Await sheathing.

P_4^c Apply plastic jacket.

Take up on shipping reel.

D_4^c Await move to inspection.

M_5^c Move shipping reel to inspection.

$_5D^c$ Await inspection.

P_5^c Inspect electrically and physically.

D_5^c Await move to shipping or stores.

M_6^c Move to shipping or stores.

Comparisons

Since there are no concurrent events or repetitive cycles in either sequence, as there were in the case of glacé fruit, total time, T, and cycle time C^*, are the same for both methods. Disregarding, for the moment, difference in yield, these values, T^b and T^c, and C^{b*} and C^{c*}, can be represented by the following summations:

$$T^b = C^{b*} = M_{1-10}^b + D_{1-9}^b + P_{1-9}^b + {}_{1-9}D^b$$

and

$$T^c = C^{c*} = M_{1-6}^c + D_{1-5}^c + P_{1-5}^c + {}_{1-5}D^c.$$

If, as seems likely, both technologies were articulated, moving times and most delay times (there were, as pointed out above, some longer delays) would have been short; nevertheless, since there were 28 such intervals in the batch process and only 16 in the continuous sequence, it is fair to conclude that

$$M_{1-10}^b + D_{1-9}^b + {}_{1-9}D^b > M_{1-6}^c + D_{1-5}^c + {}_{1-5}D^c.$$

If we then give recognition to yield difference, and multiply the left side of the inequality by a factor x that is greater than 1 but less than 10, the difference between the two technologies for these three elemental times becomes much greater:

$$x(M^b_{1-10} + D^b_{1-9} + {}_{1-9}D^b) \gg M^c_{1-6} + D^c_{1-5} + {}_{1-5}D^c.$$

The critical values, of course, are those for P^b and P^c. There are 9 steps listed for P^b and only 5 for P^c. Again disregarding the yield difference, the smaller number suggests but does not guarantee that $P^b_{1-9} > P^c_{1-5}$, because P^b might accomplish the 9 tasks in less time than P^c takes for 5. At the very least, however, parity or better for P^c was likely. If so, the yield difference would give P^c a considerable advantage indeed.

Consequently, it is reasonable to conclude that $T^b \gg T^c$, however small the yield factor may be.

By the same token, comparative values for operational efficiency, E_p, system efficiency, E_s, working-capital requirements, W, and in-process inventory, V, would show analogous advantages for the continuous technology:

$$E^c_p \gg E^b_p$$

$$E^c_s \gg E^b_s$$

$$W^c \ll W^b$$

$$V^c \ll V^b$$

A similar advantage might also be shown for floor-space requirements A^c, but in the absence of knowledge of machine and equipment size differences, this is too uncertain to claim.

In the case of the telephone cable, then, fixed-capital requirements probably were of greater significance than the modest capital sum required for the continuous processing of glacé fruit. However, for the degrees of difference shown, the necessary capital investment undoubtedly was warranted.[8]

---

8. Conceivably, the continuous process described here could have been further refined to provide continuity for the entire sequence, from the drawing of rod to storage. Capital investment for a completely continuous operation would have been high. Whether or not such an investment would have been warranted for uses that are undergoing very rapid change is not known.

TEN

Automated Systems

The word *automation* was coined in 1947 by Del Harder of the Ford Motor Company to describe the "automatic handling of parts between production processes." John Diebold in 1952 declared the word to mean "both automatic operation and the process of making things automatic." A third statement from an anonymous source defines automation as "the automatic recognition, evolution, and solution of a processing problem."[1]

My own definition is less circular than these. Automation, or, synonymously, an automated system, is one in which machines in balanced sequence perform all processes and all moves, to the exclusion of participation by humans. In "perfect" automated systems the substitution of capital for labor is made complete. In evolutionary context automated systems may be regarded as the phylogenic progeny of balanced systems.

This definition, as will be seen, applies also to programmed systems, but with a difference: in current usage automated systems are typified by "hardware," in contrast to "software," which typifies programming.

It is difficult to imagine an operations technology devoid of human content, and, indeed, no such "perfectly" automated systems exist. There are, however, subsystems in which machines perform all the tasks of processing and moving, with human intervention limited to the provision of managerial, maintenance and repair, and logistical services.

---

1. E. M. Grabbe, "The Language of Automation," in *Automation in Business and Industry*, ed. E. M. Grabbe (New York: John Wiley & Sons, 1957), p. 20.

BALANCING

As in the case of balanced systems employing humans, achievement of automation is not easy. Balance between work centers and work stations manned by human operators is complicated, as said before, by differences in performance: by diligence or indolence, by differences in aptitude, by fatigue, or by occasional product defects. These threats to system stoppages can be alleviated by changes in worker endeavor or, as has been shown, by the provision of redundant roving operators capable of relieving bottlenecks.

The machines of automated systems are not affected by these human frailties. Machines neither get tired nor relax vigilance; machines do not speed up when confronting overload, or slow down to accommodate slack. Once set, machine processing and intercenter moving times can be counted upon to be the same at the end of the working day as at the beginning, the same at night as during the day, the same in January as in June. For practical purposes automated design criteria have the inhuman virtue of consistency.

Requisite balance at, between, and among component work centers can be achieved in these ways:

1. By adjusting machine processing speeds and numbers of work stations at work centers so that the quotients of work-center processing times divided by the number of work stations are approximately equal: $(P_1/x \approx P_2/y \approx \ldots P_j/z,$ where x, y, and z are the numbers of work stations at their respective work centers).

2. By adjusting, in combination with the above, moving times and distances to achieve equality between component work centers: $[(P_1/x + M_1) = (P_2/y + M_2) = \ldots (P_j/z + M_j)]$.

3. By overdesign. If, at one or more work centers, processing and moving intervals and numbers of work stations cannot achieve requisite consonance with other system centers, additional work stations may be provided, subject to the requirement that the added cost of capital investment and sequencing idleness will be less than the expected costs of system stoppage. Thus, if Work Center 2 is out of synchrony, it may be brought into satisfactory balance by augmentation with an additional work station so that $(P_1/x + M_1) = (P_2/(y + 1) + M_2 + I_2) = \ldots (P_j/z + M_j)$.

4. Redundancy of this kind, designed to achieve intercenter balance by tolerable work-station investment and idleness, may be augmented by design redundancies that provide stand-by machines to prevent whole system stoppages. This kind of planned fail-safe redundancy is also subject to the restriction that the capital costs of redundancy and expected work-station idleness shall be less than the expected costs of whole system stoppages.

MOVING

Moving work into, out of, and between the work centers and work stations of automated systems is, by definition, accomplished by machines. Movement of units and batches of products and services by other than human conveyance has been a reality for a very long time, ever since primitive man floated a log down a flowing stream, or domesticated an ox or horse to pull or carry burdens too big to handle otherwise. But neither these nor subsequent uses of power-driven mechanisms depict the essence of automation. Operational system movements that are originated or terminated by human intervention, whatever the means of intercenter conveyance, are *not* automated. Automation requires that the system itself provide not only the *means* of movement but the *signals* by which movements are initiated ánd, after conveyance, halted, with work in position for the next operation.

Generation of these necessary signals also was accomplished long ago, by the vanes of windmills that kept the rotating blades fronted to the wind; in clocks that embodied signals by which the striking of the hours—the movement of the striker against the bell—would be started and, at the right moment, stopped, to resume an hour later; in the whirling weights of flyball governors on reciprocating steam engines, weights that curbed excessive speeds by "flying" outward and upward to reduce the movement of steam to the cylinder and slow down the engine, then opening the throttle to admit more steam as the engine slowed and the flyballs descended (Fig. 10.1).

CONTROLLING

Each of these devices for controlling movement operated by sensing a system need: to rotate the windmill about its vertical axis and

Fig. 10.1 Watt's flyball governor. As the speed of the steam engine increases, faster rotation of the governor causes the balls to rise, lifting the lever and linkage, which in turn increasingly close the valve that regulates the flow of steam to the engine. This checks the speed of the engine, the governor slows down as well, and the flyballs assume a lower position, which in turn increases the supply of steam, and so forth. Reproduced, by permission, from Andrew Bluemle, *Automation* (New York: William Collins + World Publishing Co., Inc., 1963), p. 23.

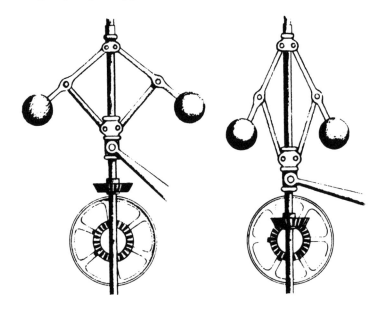

stop rotation when the blades opposed the wind at a right angle; to start the striker each hour, control the number tolled, then stop the striker; to slow and speed the steam engine within predetermined limits. Each derived the energy to exercise control by extracting ("feeding back") energy from the primary driving forces of the systems themselves: the wind, the driving force that moved the minute and hour hands of the clock, the steam that drove the engine's piston.

These and countless other control devices are capable of sensing and responding to changes in the state of the system to be controlled: changes in the direction of the wind, passage of time, and speed of the engine. For these, the primary energy sources, from which fractions are fed back for purposes of control, are, respectively, meteorological, mechanical, and thermal. Other feedback controls are capable of sensing and responding to changes in light, position, color, size, pressure, weight, viscosity, turbidity, sound, voltage, frequency, radioactivity—to almost any differences one can name.

To exercise control by acting upon properties such as these, five requirements are essential to *every* kind of control:

1. Establish limits between which a system's operation is considered to be satisfactory, beyond either of which the system's operation is considered to be unsatisfactory—that is, out of control and requiring correction.

2. Provide the capability to sense or measure whether the system is operating in or out of control, between or beyond either of the predetermined limits.

3. Provide the capability to initiate action to correct out-of-control operation.

4. Provide the capability to sense or measure when control has been restored, in order that corrective action may cease.[2]

5. Repeat items 2, 3, and 4 as necessary.

These strictures apply to *all* controls—from homeostasis, which keeps body temperature close to normal,[3] to threading a needle or steering a boat, to controlling the orbital path of a space vehicle. All the controls of automated systems act in obedience to these requirements.

Because of ubiquity, these control requisites may appear to be seductively simple, but they are not. If made too small, the tolerance limits of satisfactory operation can cause an operation or a system to "hunt," everlastingly seeking to restore control between limits set needlessly close (or fallaciously set at zero), causing whatever the corrective device may be to be repetitively on and off, first this way, then that.[4] Comparably insidious are tendencies to cause corrective actions to "overshoot," in the manner of a helmsman who perceives that he is off course to port and puts the wheel hard over, only to veer off course just as far to starboard. Hunting results in repetitious, niggling corrections; overshooting causes less frequent, wide oscillations.

2. Gordon S. Brown and Donald P. Campbell, "Control Systems," in *Automatic Control*, comp. Editors of Scientific American (New York: Simon & Schuster, 1955), p. 27.

3. "The term 'homeostasis' was coined by Harvard University physiologist Walter B. Cannon to describe the many delicate biological mechanisms that detect slight changes of temperature or chemical state within the body, and compensate for them by producing equal and opposite changes" (quoted, by permission, from W. Gray Walter, "An Imitation of Life," ibid., p. 125).

4. Needlessly small tolerances in automated controls, quality control, and risk assessment have cost society dearly. Exacerbation has come from evangelistic but fallacious "zero defects" crusades in industry and comparably fervent pursuit of "zero risk" from nuclear power plants and from waste disposal.

To be successful, an automated system, being capital intensive, must operate in control most of the time. Designers of automated systems must achieve operational continuity for whatever the whole system may be, which means that controls for each process and each move must be reliable, lest failure in any single interdependent component cause system stoppage. This is a formidable assignment, made even more difficult by decreasingly small control limits imposed by process necessities.

Fortunately, developments in control theory and technological progression have made possible sensing devices capable of responding to very small differences, actuating controls of great delicacy and high reliability.

Some idea of this progression can be exemplified. In the ancient "automated" windmill, sophisticated theory and technology were unnecessary. If the windmill's blades were fronting toward the west and the wind veered to the north, the force on the windward side of the vane would rotate the blades about the mill's vertical axis to front north. As the mill approached the desired position, the rotational force component would approach zero, and any tendency to overshoot would be self-corrective, for what had been the leeward side of the vane would now become the opposite, windward side.

The flyball governor, not so old as the windmill and vane, but still a vintage control device, tells a different story. Fuel with which to generate steam is not free like the wind. Thermal efficiency, which depends upon optimal engine speed, becomes important. Design of the governor itself requires attention not only to feasible fast-slow control limits but also to the dynamics of the governor itself: the weight and number of balls, their rotational speed, and linkage with the steam valve to modulate the flow of steam to the engine.

Neither of these automated machines uses electricity for purposes of control, but most machines do. Differences in system status can be detected or measured for properties of diverse kinds, as said before, but control responses are very likely to be electrical, or, more recently and more frequently, electronic.

Years before the word *automation* was coined, "transfer machines" (Fig. 10.2) were created to perform half a hundred consecutive metal-working operations and as many interoperational moves—with human intervention limited to supply and monitoring. On a smaller scale, automatic screw machines, fed with metal rod of appropriate diameter, could do the same kinds of things, with single operators attending batteries of machines, monitoring stoppages, replacing dulled cutting tools, and supplying raw material.

In the textile industry, invention of the Draper loom by J. H.

Fig. 10.2 Transfer machine for the automatic performance of multiple-sequence operations. Courtesy of Avey Machine Tool Company.

Northrop (Fig. 10.3) made it possible, without stopping the loom, to insert "cops" of new weft thread into the shuttles that traveled back and forth between the alternating up and down threads of the warp.[5] One weaver could attend as many as 24 Draper looms, maintaining stocks of new cops and monitoring stoppages occasionally caused by broken thread. So reliable were these machines that weavers, who were paid "pick rate" (wages paid in proportion to the count of shuttle passes across the loom), could let their looms continue to operate during lunch periods, in order to earn extra pay. When resuming work weavers restarted the looms that had stopped while unattended.[6]

So far, we have described automatic, self-contained machines: windmills, clocks, steam engines, transfer machines, and looms.

5. D. A. Farnie, "The Textile Industry: Woven Fabrics," in *A History of Technology*, ed. Charles Singer, E. J. Holmyard, A. R. Hall, and Trevor I. Williams (London: Oxford University Press, 1954), 5:585.

6. This practice was revealed during an arbitration hearing at a silk mill, by report of loom efficiencies, computed by the ratio of actual picks made to the total number of possible picks. Seemingly impossible efficiencies exceeding 100 percent were reported. When asked how this could be, weavers testified that their loom sets were allowed to continue to run while the weavers ate lunch.

Fig. 10.3 Draper automatic loom hopper mechanism. Insertion into the shuttle of a fresh cop of weft thread ejects the empty cop and enables the loom to continue weaving without stopping. Reproduced, by permission, from D. A. Farnie, "The Textile Industry," in *A History of Technology*, ed. Charles Singer, E. J. Holmyard, A. R. Hall, and Trevor I. Williams (London: Oxford University Press, 1958), 5:585.

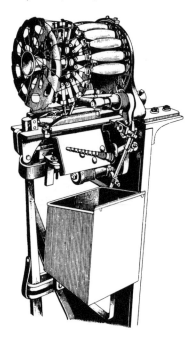

These are not automated *systems* in the sense of current usage, but they do embody the essence of automation. In a transfer machine, completion of a cut, sensed by the position of the part, actuates return of the tool to its starting position and engages it with the next piece, coincidentally placed in position. At the same time, the piece that has been cut moves into position for the next operation and is there engaged by whatever the successor tool may be. Analogously, exhaustion of the thread in the cop carried by the shuttle signals need for a replacement, and the empty cop is ejected and replaced by a new one during the instant the shuttle is at rest.

The sensory signals for these controls vary, but the response actions depend upon the quickness, reliability, and delicacy made possible by electricity. Electricity continues to be the kind of energy used to power—and often to sense—control responses, but electronics has refined these powers, one may say to the nth degree, to make possible control limits that are almost incredibly small.

For example, large web-fed, multicolor printing presses can produce precisely registered color illustrations at very high speeds. Successive impressions of yellow, red, blue, and black ink are superimposed in such a way as to avoid "out-of-register" blurring perceptible to the unaided eye. Control is exercised by a photoelectric sensor that "looks at" the edge of the paper as it unreels from the roll and speeds through the machine. Very small changes in edge position are detected by the photocell and fed back to a servo-mechanism, an electric motor that rotates just enough to change tension on the web by moving the angle of the roller around which the paper passes. An observer can see the servo move now and again as it exercises control, but variations in register remain too small to be perceptible.[7]

Perhaps the ultimate in control refinement will be reached in space technology. Already realized are vernier rockets, which provide directional control over launch vehicles by firing with just enough thrust to keep propulsion from the main engines on course, and do so within very close limits. Down the road lies a more difficult problem, that of aiming the space telescope at some very distant star. This perhaps ultimate problem has been declared by one astrophysicist to be about as difficult as standing in Baltimore, aiming at a dime held in Boston, and hitting the target in the middle.[8]

These advances—the extraordinary sensing and controlling capabilities of electronics—have made possible the extension from automated machines to automated systems: the coupling of one transfer machine to another engaged in different kinds of tasks; the coupling of printing and slicing the web into sheets for folding, collating, wire stitching, trimming, addressing, and mailing mass-circulation magazines; the washing, sterilizing, and drying of empty

7. Within the recent past, daily newspapers have increasingly used color, and possibly because of the less consistent quality of the paper used (newsprint), have been experiencing problems in controlling color register. The coincidental change from reproduction by letterpress to offset lithography may also have contributed to this difficulty.

8. This metaphor came in response to an audience question asked of Dr. Arthur Davidsen, Professor of Astrophysics, Johns Hopkins University. From another colleague, Professor William G. Fastie, I have learned that in searching for, finding, and fixing upon an unknown, very distant star, the space telescope will orient upon nearer stars of known position. In scanning, the telescope will be sensitive to light energy (feedback) as small as 10^{-23} watt, and function with angular control limits of approximately (if one may use such a word for a dimension so small) 7 milliarc seconds (rms).

Fig. 10.4 Machines for automatically blending, cooling, carbonating, filling, and crowning bottled soft drinks. An additional machine for automatically washing, sterilizing, and drying empty bottles is not shown but can be made part of the sequence. Courtesy of Archie Ladewig Company, Division of Crown Cork and Seal Company, Inc., Baltimore, Md.

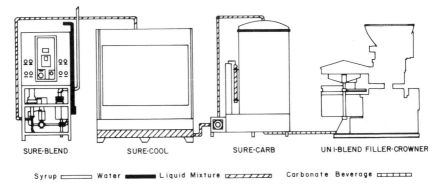

bottles, followed by the blending, cooling, carbonating, filling, and crowning of bottled soft drinks (Fig. 10.4), all in automated sequences.

ANALOGS

A pantograph is a tool with which, by suitable linkage, a stylus traces a path on a model, or analog, while some form of cutting tool follows a like path on a given raw material, enlarging or reducing from the analog according to some desired proportion.

To make a steel punch for the roman capital letter A of this text, for example, the pantograph stylus can be guided around an analog of a much larger A and around the triangular counter at the top of the letter, while the cutting tool removes metal to form the raised image of the letter, faithful to, but much smaller than, the original design embodied in the analog. Smaller or larger A's can be made by changing the linkage to different proportions. More to the point here, by simply changing analogs to large models of other letters and other typefaces (a, *B*, **C**, 8, etc.), the same machine and tool can be made to do many *different* things.

The above process is one in which the operator guides the stylus by hand, but the procedure admits of automation: the analog can be contrived to guide the movement of the stylus (Fig. 10.5) while the

Fig. 10.5 Contouring machine in which an input stylus (toward the outer end of the arm to the right) follows the contours of an analog to guide the cutter (shown over the blades in the center) of the machine. By changing patterns, one can produce a variety of shapes. Courtesy of George Gorton Machine Company.

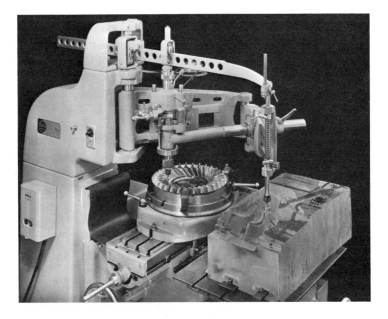

tool cuts the raw material into the intricate shape of the model.

Earlier, I described jobbing systems as having many disadvantages but one salient advantage: the capability to do different things. I then discussed how articulation, balance, and continuity overcame the disadvantages of jobbing, but did so by trading them, so to speak, for the restriction of sameness. Here we can see that through the use of analogs, automation has taken a step toward recapturing capability for differences.

In a not too fanciful sense, one can say that transfer machines, automatic screw machines, and Draper looms have been *programmed* to perform their designed sequences of operations and moves. This is also true of the automated pantograph, but with a difference: the pantograph's program is the analog, and the analog can be quickly and easily changed to guide as many different shapes, sizes, and kinds as may be desired.

Other machines also have embodied this capability for "easy differences": Jacquard looms, Monotype paper tapes, the punched cards of Hollerith. These and other programs like them have been harbingers of the next evolutionary step in operations technology: *programmed systems*, the subject of Chapters 12, 13, and 14.

SYSTEM STOPPAGES

As long as automated systems operate without human intervention, the time dimensions specified in the equations of Chapter 1 are less meaningful than those for predecessor systems. With automation, process time, P, is short and devoid of wage cost; moves, M, are mechanized and of brief duration; and delays, D, and sequencing delays, I, are zero. At the same time, automated system output, S, is likely to be larger than outputs from predecessor system technologies.

As a consequence of these advantages, in automated systems total time, T, and cycle time, C^*, are short. From high output values, S, in numerators and low input values in denominators, operational efficiency, E_p, and system efficiency, E_s, are high. Working capital, W, in-process inventory, V, and floor-space requirements, A, are all relatively small.

These advantages are real enough for automation while such systems are operating. But being capital intensive, such systems demand high load factors: they must operate most of the time. It is important therefore to include holding time, H, when comparing automation with other system technologies. We have no means for assigning a value for H but can exemplify its effect.

A nuclear power plant is essentially an automated system for generating electricity, with human intervention serving to monitor rather than operate. As long as such a plant is running, its output, S, will be high and the denominators of operational and system efficiencies, E_p and E_s, will be small. For an operating nuclear power plant the cost per kilowatt-hour is lower than can be attained by almost any other means of generation.

Eventually, however, the fissile fuel requires replacement, necessitating a lengthy holding interval, H. During this shut-down no electricity—nor any revenue—is generated by the facility, but the carrying costs of the capital-intensive plant continue. When these charges are factored into the cost per kilowatt-hour ratio, the effect of the stoppage is to raise the quotient by a significant, sometimes substantial, amount.[9]

This example may not typify the effect of stoppages upon automated systems in general, but one can say that capital-intensive systems that consist of interdependent components are more vulnerable to stoppages than any other technological form. At the extreme, malfunction of *any* component will cause the entire system to stop. The cost of the stoppage will be directly proportional to carrying cost for whatever the duration of H may be.

9. Opponents of nuclear power often disagree with rate calculations made by utilities, claiming that load factors are unrealistically high—never attained or not likely to be achieved. On more than a few occasions, load factors used by those who disagree are sufficiently lower to yield an apparently higher rate. The difference lies in the value chosen for H.

ELEVEN

Automation within a System

THE MANUFACTURE OF
DRY CELL BATTERIES

Transformation from an earlier articulated system to one that was partially but significantly automated took place in India during the years following World War II, when the influence of Gandhi waxed while that of the British raj waned, when newly won independence confronted enterprise managers with new and formidable restraints: blocked currency, scarce capital, and labor unions capable of militant opposition. These and the spirit of "Small Is Beautiful," personified by Gandhi and extolled by Schumacher,[1] required a different calculus for decisions about technological change.

Dry cells[2] are made in various shapes and sizes, but those to be described here are small cylinders consisting of four principal parts, as shown schematically in Fig. 11.1 and listed below:

A carbon anode (the positive electrode)
A "depolarizing mixture," consisting of manganese dioxide, highly conductive acetylene black, ammonium chloride, and zinc chloride
An electrolyte consisting of ammonium chloride, zinc chloride, and wheat flour
A container ("can") made of zinc-cadmium alloy, which forms the cathode (negative electrode) of each cell

1. E. F. Schumacher, *Small Is Beautiful* (New York: Harper & Row, 1973).
2. For this account I am very much indebted to my nephew, Mr. J. Roy Galloway, now retired from his former position as Vice-Chairman of Union Carbide Eastern. At the time of the operational methods described, Mr. Galloway was Managing Director and Chairman of the Board of National Carbon Company India, Ltd., with headquarters in Calcutta.

Fig. 11.1 Schematic drawing of the principal components of a dry cell battery.

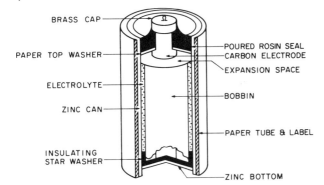

Other parts will be introduced as necessary in the process descriptions.

THE PREDECESSOR ARTICULATED METHOD

Carbon anodes were first extruded in rod form, then baked in a gas-fired furnace to make the rods permeable and hard. This was followed by centerless grinding and sawing to specified length on diamond saws. Precision dimensions for diameter and length were required. Finished anodes were placed in boxes and held for assembly with the depolarizing mixture.

To make the depolarizing mixture, the four ingredients in proper proportions were put into a rubber-lined rotary machine and mixed to a doughlike consistency. Following wetness tests to ensure proper moisture content, the mix was moved on wheeled "buggies" to the "bobbin department" ("bobbin" was the name applied to the assembly of anode and depolarizing mix). Because the material was unusable if it dried out, prompt utilization in assembly was necessary. Upon arrival in the bobbin department, the mixture was transferred by a move man into a hopper, from which it was taken by one of the two operators (a "pusher" and a "catcher") who manned the four-stage device that created the finished bobbins. This device, schematically suggested in Fig. 11.2, consisted of a round table that was manually rotated by the pusher. There being four steps in the assembly of each bobbin, each rotation step took one-quarter turn.

In the round table, 90° apart, were four holes, in each of which

Fig. 11.2 Four-position device for assembling bobbins.

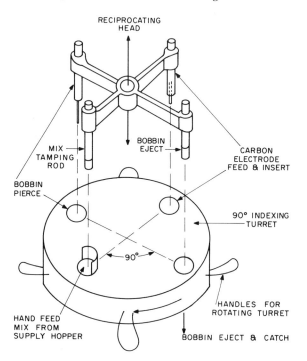

was a stellite[3] insert of inside diameter precisely equal to the outside diameter of the bobbins to be assembled. For purposes of discussion, let us assume that these four positions are numbered 1, 2, 3, and 4, in clockwise order.

At position 1, the pusher, with his left hand and with remarkable nicety, took a quantity of mix from the hopper and put it into insert 1. The pusher then depressed a foot treadle that simultaneously tamped the mix just placed in insert 1, pierced a center hole in the previously tamped mix in insert 2, inserted a carbon anode in the previously tamped and pierced mix in insert 3, and ejected the previously finished bobbin now in position 4. The catcher caught the ejected bobbins between his fingers, inspected them, and put them into trays with cardboard separators. Trays of finished bobbins were then moved to the "cooking department," there to be joined with cans and electrolyte.

In a separate sequence, cans were made from coils of zinc-

3. Stellite is a very hard and wear-resistant alloy.

Fig. 11.3 Schematic diagram of a soldering table.

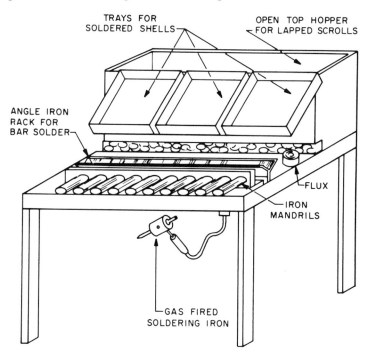

cadmium alloy of correct thickness (gauge) and of width equal to the height of the can. (Cadmium serves to harden zinc and inhibit corrosion and perforation of the can's wall.) Coils were then unreeled into a "kick press" by an operator who cut them to correct length. These flat pieces were stacked in trays and moved to the "scroller."

The scroller was a hand-fed machine that rolled each piece into a cylinder of slightly smaller diameter than that of a finished can, so that there was a small lap at the seam of each now-cylindrical piece. These were placed in a "tea chest" for transportation to bins in the can department.

To form these lapped scrolls into soldered, bottomless cylinders, two operators worked at adjacent tables, as suggested by the schematic drawing in Fig. 11.3. Along the front edge of each of the two tables were ten mandrels, each cylindrical and each of outside diameter equal to the specified inside diameter of a can. Bins of lapped cylinders were located within easy reach of each table, at which there were also supplies of flux and solder, and trays to contain finished cylinders.

To visualize the cycle of work, let us assume that at the start there were ten cylinders on the mandrels of table 1 that had been placed there seam-up by operator A, who had also brushed each seam with soldering flux. On table 2 there also were ten cylinders, the seams of which had been soldered by operator B.

At this point operators A and B exchanged positions, B going to table 1 and A to table 2.

On arriving at table 2, operator A wiped off the ten soldered seams, then removed the ten cylinders from the mandrels and placed them in a tray. He then placed ten unsoldered cylinders seam-up on the mandrels of table 2, applied flux to each seam, and prepared to move back to table 1.

During this interval, operator B soldered the flux-prepared seams of the ten cylinders on the mandrels of table 1 and, by going to table 2, again changed places with operator A.

Each such pair of operators functioned as a team; they worked, necessarily in balance, on piece rate and could exchange tasks when and if they so desired. The trays into which these soldered, bottomless cylinders were placed were then put on skids and moved to the bottom-soldering department.[4]

Shaped and embossed bottoms were purchased. These were stacked in uniform position in a feeder tube, from which they were delivered, one at a time, to a fixture into which the bottom fit precisely. As this was done, the operator placed a bottomless cylinder on a hinged mandrel and swung the mandrel to a vertical position so that the bottom was fitted into the cylinder to just the right depth. Cylinders were then placed bottom-up in trays and moved to soldering.

Two men participated in soldering: one applied flux and the other actuated soldering by depressing a foot treadle that applied solder from a pot of molten metal. These operators, also on piece work, were allowed to change positions. Cans, now complete with soldered seams and bottoms, were placed in trays and moved to inspection. Inspection was visual and cans that were perceived to have leaks were rejected. Finished and inspected cans were then moved to the "cooking department," there to be assembled with electrolyte and bobbins.

Each can that passed visual inspection was put upright into a

---

4. At the Singapore plant an improved method was devised by mounting mandrels on a rotating wheel. The operational sequence was the same and two operators were required, but they no longer had to change places.

"cooking tray" divided into ten ten-can rows. Each such "indexed" tray was then moved to a "paste machine," where an operator moved the first row under a nozzle that measured an exact amount of electrolyte into each can. This was repeated for the remaining nine rows, after which the cooking tray was moved to a "star washer" machine.

There an operator inserted a wax-impregnated cardboard star washer into the top of each can. The trays then moved into the "cooker," a hot-water tank through which cans were conveyed to keep the gelatinous electrolyte sufficiently fluid to flow up and around the depolarizing mix as each bobbin was inserted (see Fig. 11.1). The task of insertion was performed by eight operators, four working on each of the two sides of the cooker. Each operator was supplied with bobbins to be inserted into cans now containing electrolyte and star washer. The work of forcing down the star washer with the bobbin required delicacy and precision, to ensure that the electrolyte was forced up to just the right height: if forced too high, electrolyte would spill over the top of the bobbin and come into contact with the anode, thereby causing a short circuit; insufficient height would diminish the cell's electrical output. It was also important that the anode be centered and parallel to the inside of each can.

Emerging from the cooker, the gelatinous electrolyte set as it cooled and the assembled cells, still in the indexed trays, were moved to a kick press, where the wax-impregnated washers were inserted into the tops of ten cells at a time, in such a way as to leave space above the bobbins for gases (see Fig. 11.1). Each batch of ten cells was ejected onto flat, open-ended metal trays. These were next taken to a foot-operated machine that applied a brass cap to the protruding top of each anode.[5]

Cells were then placed in trays and moved to a work station where their tops were sealed with rosin poured from a special ladle in such a way as to form a tiny miniscus around the inside circumference of the can and the outer circumference of the anode.

The trays were then placed in storage for one week to detect short circuits, leaks, raised electrodes, and run-down cells. All cells were then tested with a voltmeter before movement to a tubing department for application of a protective tube and decorative lable, after which the now-completed cells were packaged for shipment.

---

5. Brass caps were more decorative than functional. They are no longer used.

AUTOMATION

Only two stages of this labor-intensive system were automated.

The less significant of these involved the making of bobbins. For this part of the system, the four-stage, rotating table shown in Fig. 11.2 was still used, as were the stages of tamping, piercing, inserting the anode, and ejecting the finished bobbin. However, these operations and the rotation that followed were made automatic, so that a single operator, now machine paced, could remove finished bobbins and place depolarizing mix into the stellite insert in synchrony with each quarter-turn of the machine.

For this stage, output was increased with but half the labor, at modest capital cost.

Automation of can manufacture was much more significant, at greater—but still modest—capital investment.

All the moves and processes, from zinc-cadmium rolls to finished cans, were made automatic in such a way that multiple-machine operation became possible: one operator could attend four of the machine sequences diagrammed in Fig. 11.4 and described below.

A coil of alloy of width equal to the height of the can and of suitable gauge was fed through rolls and cut into flat sheets. These were moved through successive steps by a reciprocating mechanism with teeth that moved each piece forward, first to be formed into a lapped cylinder, but this time with seam down rather than up, as in the earlier method. This underside seam then moved over a wheel that applied flux, after which the seam moved ahead over a wheel rotating in a bath of molten solder.

When cooled, the soldered cylinder, still bottomless, was thrust upon a mandrel, one of a series mounted on the circumference of a rotating wheel, each such mandrel synchronized to receive each successive cylinder after soldering and cooling.

As these mandrels rotated about the axle of the wheel upon which they were mounted, they were also turned so that each cylinder, after insertion of its bottom from feed stock, would rotate to receive flux and solder around the circumference of the can.

Finished cans were then ejected upon a conveyor belt and removed by an operator using a kind of multitined fork to put the finished cans into trays for moving to inspection and thence to the cooking department.[6]

6. At a later time, a rolling mill in the United States was disassembled, shipped to India, and there reassembled. The mill was used to roll the zinc-cadmium alloy,

Fig. 11.4   Schematic diagram of automated can making.

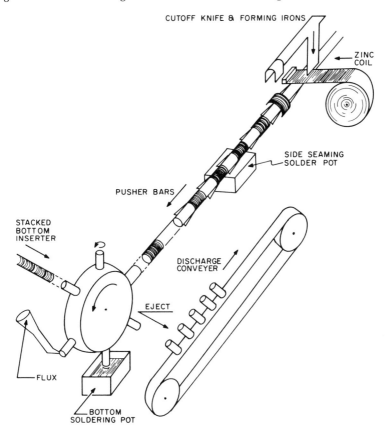

COMPARISON

The system changes that have been described did not affect the production of carbon anodes; the insertion of electrolyte, star washers, and bobbins; the application of wax-impregnated washers and brass caps; sealing with rosin; or storage, testing, application of protective tubes and labels, and packaging for shipment.

For the making of bobbins, there was no change in the four-stage sequence, but the "pusher" was relieved of the task of manually rotating the table by the installation of a power drive so timed as

and from its output, cans were extruded as seamless units, to the exclusion of all soldering.

Table 11.1 Flow process chart for articulated production of cans

P_1^b	Cut length of zinc-cadmium alloy from roll. Stack on tray.
D_1^b	Await move to scroller.
M_2^b	Move tray to scroller.
$_2D^b$	Await scrolling.
P_2^b	Form flat piece into lapped cylinder. Place in tea chest.
D_2^b	Await move to can department.
M_3^b	Move tea chest to can department.
$_3D^b$	Await soldering operations.
P_3^b	Place ten cylinders seam-up on mandrels.
P_4^b	Apply flux to ten cylinders.
P_5^b	Solder ten cylinders.
P_6^b	Wipe ten soldered seams.
P_7^b	Remove ten soldered cylinders. Place on trays.
D_7^b	Await move to bottom soldering department.
M_8^b	Move trays to bottom soldering department.
$_8D^b$	Await insertion of bottom.
P_8^b	Insert bottom to proper depth. Place bottom-up in tray.
D_8^b	Await move to bottom soldering.
M_9^b	Move to bottom soldering.
$_9D^b$	Await bottom soldering.
P_9^b	Apply flux.
P_{10}^b	Solder bottom to cylinder. Place in tray.
D_{10}^b	Await move to inspection station.

NOTE: Operations P_{3-7}^b were performed by two workers, as described in text.

to make possible operation by one worker instead of two, and there was an additional gain from machine pacing.

These were not inconsiderable gains, but that from the automation of can production was far more dramatic.

Table 11.1 is a flow process chart of the articulated sequence, denoted by the superscript *b* to indicate that processing delays and moves signify batches of cans rather than the one-by-one method of Table 11.2, which is denoted by the superscript *a* for automation.

There being no detailed data for these time intervals, nor batch sizes in the articulated sequence, exact comparisons are not possible. Nevertheless, assuming that each chart portrays the passage of a single can through each system technology, meaningful differences can be distinguished.

Table 11.2   Flow process chart for automated production of cans

P_1^a	Cut length of zinc-cadmium alloy from roll.
M_2^a	Move by machine to reciprocating conveyor.
P_2^a	Form into lapped cylinder, seam-side down.
M_3^a	Move to flux wheel.
P_3^a	Apply flux to seam.
M_4^a	Move to solder wheel.
P_4^a	Apply solder to seam.
M_5^a	Move to rotating wheel with mandrels.
P_5^a	Place cylinder on mandrel.
M_6^a	Move to application of bottom.
P_6^a	Insert bottom to proper depth.
M_7^a	Move to application of flux.
P_7^a	Apply flux to bottom on rotating mandrel.
M_8^a	Move to soldering.
P_8^a	Solder bottom to cylinder on rotating mandrel.
M_9^a	While cooling, move to ejection.
P_9^a	Eject to conveyor belt.
P_{10}^a	Remove with fork. Place on tray.
D_{11}^a	Await move to inspection station.

In the articulated sequence there were nine delay intervals, most of them interdepartmental. Since the system was articulated for the manufacture of a single product, it may be assumed that approximate balance existed among and between operations, so that these delay intervals probably were short but in the aggregate not inconsiderable.[7]

In the automated sequence, there is but one delay interval preceding the movement of finished cans to inspection. All other delays have been eliminated by the perfect balance of automation.

Unit differences in the interoperation moves involved in the two methods were probably small, but such as there were undoubtedly favored automation.

The greatest difference between the two methods derived from

7. For batches of product, each delay interval comprises two components: (1) time during which each batch (tray, tea chest, skid, etc.) awaits being moved, or awaits commencement of processing after being moved, plus (2) processing time for the entire batch. For a single unit, delay time for an "average" can will equal the sum of these intervals divided by the number of cans in the batch: $\overline{D} = (D^b + P^b)/b$.

Fig. 11.5 Annual production of dry cells, 1940–55. Current output is about 500 million cells per year.

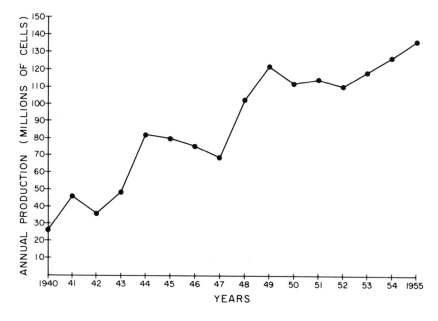

the processing superiority of automation, for which quantitative comparison can be approximated:

In the articulated system, can production for each team of solderers was about 1,500 per day.

In the automated sequence each mechanized unit produced approximately 18,000 cans per day, a 12-fold increase. Four of these machines were attended by one operator.

Thus, in terms of the equations of Chapter 1, automation has had favorable consequences: total time and cycle time have been decreased; operational and system efficiencies have been increased; and working-capital requirements, in-process inventory, and floor-space requirements have been decreased.

These conjectures are impressive but not quite conclusive. Two considerations remain: fixed-capital requirements for system innovation, and technological unemployment.

The first of these has already been addressed by discussion of the simple changes effected in the making of bobbins. Mechanization and synchronization of the rotating table required fairly simple design and inexpensive modification, carried out, no doubt, in house.

Automation of can production was considerably more difficult, but it too admitted of in-house design, prototype fabrication, and testing. This change, and even the later change to extrusion, bespoke fixed-capital conservation in an economy of blocked currency.[8]

Technological unemployment was a more sensitive and, at the same time, more interesting problem. Had the market for cells been stable, it seems doubtful that automation would have been feasible, but growth was sufficiently rapid (Fig. 11.5) to create increased employment and transfers rather than separations for those affected by the technological changes.

In the event, the changes and the market expansion *were* made. This suggests that the quest for doing more for less may appear more "beautiful" than "smallness," even in India.

8. An unusual means of acquiring capital deserves mention. Observing that Indian fishermen embarked each morning and returned each evening with catches of shrimp, Union Carbide was persuaded to invest in a trial factory ship for voyages of longer duration in the Indian Ocean. Catches were cleaned, processed, packaged, and frozen on board, and the product was exported as a means of bringing in foreign capital. There are now eleven shrimp vessels in the fleet.

TWELVE

Programmed Systems

A programmed system can be described in the same words that were used in Chapter 10 to define an automated system: both are systems "in which machines in balanced sequence perform all processes and all moves, to the exclusion of participation by human beings." In "perfect" programmed systems, just as in "perfect" automated systems, the substitution of capital for labor is made complete.

These denotative likenesses—mechanization of processes and moves and exclusion of human participation—may suggest that programming is nothing but an extension of automation, a minor mutation in the evolution of operations technology. Not so. Whether the transition between automation and programming can properly be called a "punctuational" evolutionary step, I do not know, but the impact of programming upon systems technology can be compared to the earlier effects of articulation. By trading sameness for variety, the speed and economy of articulation beneficially transformed countless jobbing systems into articulated sequences. In an even more dramatic way, information encoded in the programs of electronic computers has extended automation into many more, different, and larger system sequences, manned by fewer and fewer humans. Programming has also restored capabilities for variety to the whole domain of operations technology. If the impact of programming deserves a punctuation mark, it deserves an exclamation point!

In various ways earlier chapters have implied the existence of programs, even when the word itself has not been used. Design of an articulated, balanced, or continuous system has been said to provide control of the flow of goods and services through whatever the sequence of processes and moves may be. In such systems the designed configuration of work centers and work stations, along with requisite balance, can be called the system's program. So, for that matter, can the flexible scheduling devices of jobbing systems: machine-loading charts, order-of-work lists, or court dockets can be called programs for the time intervals projected.

Usage of this kind could also be properly applied to the designs of the flyball governor, the tolling clock bell, transfer and automatic screw machines, and Draper looms. Each has been programmed to do its appointed task.

Nearer, however, to the usage intended here are the kinds of programs that instruct the Jacquard loom, the Monotype caster, and the Hollerith sorter. Each of these machines receives instructions by means of holes: holes sensed by needles in Jacquard looms (Fig. 12.1), air escapement through holes perforated in Monotype tape (Fig. 12.2), and electrical contacts through holes punched in Hollerith cards (Fig. 12.3).

In each of these the presence of a hole allows the flow of information, telling the loom to lift a particular thread, the Monotype to cast an *A*, the sorter to send a card into the sixth pocket. In each of these—and this is equally important—the *absence* of a hole *prevents* the flow of information, telling the loom *not* to lift that particular thread, the Monotype *not* to cast an *A*, the sorter *not* to send that card into the sixth pocket. In each case the means of sensing is different—by needles, air, or electricity—but the logic is the same: a hole means *yes*, no hole means *no*.

This kind of "binary logic" has been used for a long time. As long ago as the early seventeenth century the alphabet was encoded into combinations of only two-letter symbols (Table 12.1) for purposes of secrecy. Communication at a distance was accomplished by the ringing of two bells of different pitch, and by flashing light through different configurations of horizontal and vertical slits.[1] More recent, and still familiar to all but the very young, is the Morse code, the binary dots and dashes of which helped make telegraphy possible.

Binary numbers, using only the two symbols 0 and 1, also were proposed long ago,[2] but extensive usage has been much more recent, in the "and," "or," and "not" circuits of present-day computers (Fig. 12.4). The equation $x^2 = x$ of George Boole, for which the only solutions are the digits 0 and 1,[3] Claude Shannon and Warren Weav-

1. Volker Ashoff, "The Early History of the Binary Code," *IEEE Communications Magazine*, January 1983, pp. 4–10.

2. Ibid., p. 9.

3. Herman H. Goldstine, *The Computer from Pascal to von Neumann* (Princeton: Princeton University Press, 1972), chaps. 4, 7.

Fig. 12.1 Diagram of the control mechanism of the Jacquard loom, invented in 1801. "The hooks *aa* pass perpendicularly through eyes in the needles *bc*, which are fixed in the frame *dd*. . . . The needles protrude at *c* and are kept in position by springs *ee*. Above the frame *dd* is another frame *h*, which is alternately raised and lowered. The hooks are lifted by the bars in the frame *h* if they are retained in a vertical position. If they are thrust out of this position (as shown by the four lower hooks) the bars miss them and they remain down. The pattern of the perforations in each card as it is pressed against the needles at *c* determines which needles are thrust back and which are not; that is, a perforation allows a needle to remain stationary, and its hook vertical, so that the hook and the corresponding series of warp threads are raised. A blank in the card thrusts a needle backwards, its hook remains lowered, and the warp threads controlled by the hook remain below the weft." Reproduced and quoted, by permission, from W. English, "The Textile Industry: Silk Production and Manufacture, 1750–1900," in *A History of Technology*, ed. Charles Singer, E. J. Holmyard, A. R. Hall, and Trevor I. Williams (London: Oxford University Press, 1958) 4:316–19.

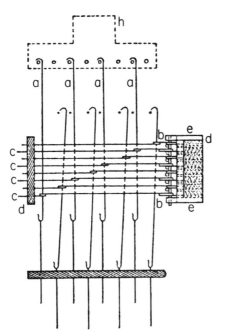

Fig. 12.2 A short length of Monotype tape. Rolls of such tape are encoded by striking keys much like those of typewriters. Each key stroke punches two holes in each successive row of the 32-channel tape, in combinations of pairs that signal positions for a matrix case (a matrix is the mold for each letter), which holds 255 characters. Completed rolls are placed on the tower of a companion casting machine. As the tape moves, row by row, over the tower, compressed air passes through each of the two holes (and does *not* pass through the other 30), to move the matrix case north or south and east or west, thereby bringing the signaled letter over the mold. The injection and solidification of molten type metal then forms the body and face of the letter, which is then ejected into the line of type.

The ribbon passes in a forward direction over the tower of the keyboard and then moves backward over the tower of the casting machine. By this means, the justifying signals, shown by the pair of larger holes near the bottom of the illustration, pass over the tower before the characters in the line itself. The justifying signals, punched by the keyboard operator at the end of each line, actuate two stepped wedges, one with step intervals of 0.0075 inch and a second with intervals of 0.0005 inch. In proper combination these add an increment of width to the 6-unit (1/3 em) fixed spaces signaled by the keyboard space bar and shown several times in the figure. Note that the keyboard operator failed to key a hyphen after *cele*. Courtesy of Herbert F. Czarnowsky, Jr., Volker Brothers, Baltimore, Md.

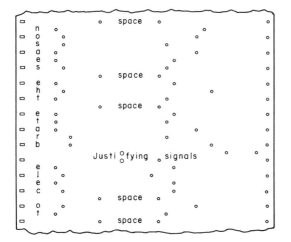

Fig. 12.3 A Hollerith punched card in which there are 80 columns, each with 12 rows. Stacks of these cards are placed in a sorting machine, where each card is passed over an electrical contact that permits a signal to flow through a punched hole (at the same time *not* permitting signals to pass elsewhere). If the sorter were set to scan column 3, this card, and all others with a hole in row 6, column 3, would fall into pocket 6, while cards punched for other digits in column 3 would fall into their respective pockets.

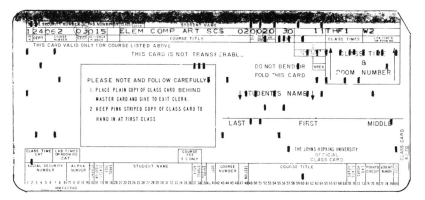

Table 12.1 Francis Bacon's two-letter alphabet of 1605

A	=	a a a a a		H	=	a a b b b
B	=	a a a a b		I	=	a b a a a
C	=	a a a b a		K	=	a b a a b
D	=	a a a b b		L	=	a b a b a
E	=	a a b a a		M	=	a b a b b
F	=	a a b a b		N	=	a b b a a
G	=	a a b b a		O	=	a b b a b

SOURCE: Adapted from Volker Ashoff, "The Early History of the Binary Code," *IEEE Communications Magazine*, January 1983, p. 5.

er's *Mathematical Theory of Communication*,[4] and John von Neumann's proposal to change from the decimal numbers used by ENIAC to the binary numbers of IAS[5] all contributed to the pervasive use of binary numbers by digital computers (Table 12.2).

To say this is to jump ahead, however. Computation by machines was achieved long ago, by Pascal, Leibniz, and others. Boole's

4. Claude E. Shannon and Warren Weaver, *The Mathematical Theory of Communication* (Urbana: University of Illinois Press, 1949).

5. The proposal was made in an unpublished but influential paper entitled "First Draft of a Report on the EDVAC" (Goldstine, p. 188).

Fig. 12.4 "And," "or," and "not" circuits using (a) switches actuated by relays and (b) semiconductors and transistors. In the circuits at the upper left and right there will be an output only when there are inputs on the top *and* bottom leads. In the circuits at center left and right there will be an output if there is an input on either top *or* bottom lead. In the pair of circuits at the lower left and right an input will *not* produce an output, while the absence of an input will produce an output. The last of these arrangements is called an "inverter." Reproduced, by permission, from Willis C. Gore, "Electronic Digital Computers," in *Operations Research and Systems Engineering*, ed. Charles D. Flagle, William H. Huggins, and Robert H. Roy (Baltimore: Johns Hopkins Press, 1960), p. 290.

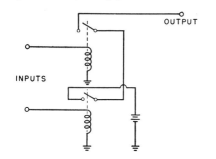

RELAY "AND" CIRCUIT

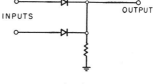

DIODE "AND" CIRCUIT

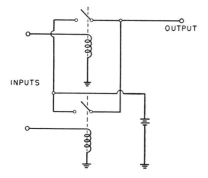

RELAY "OR" CIRCUIT

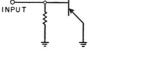

DIODE "OR" CIRCUIT

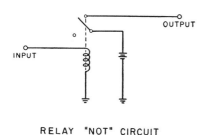

RELAY "NOT" CIRCUIT

(a)

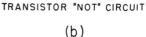

TRANSISTOR "NOT" CIRCUIT

(b)

Table 12.2   Binary arithmetic

				Binary counting			
Decimal	*Binary*	*Decimal*	*Binary*	*Decimal*	*Binary*	*Decimal*	*Binary*
0	0	6	110	11	1011	16	10000
1	1	7	111	12	1100	17	10001
2	10	8	1000	13	1101	18	10010
3	11	9	1001	14	1110	19	10011
4	100	10	1010	15	1111	20	10100
5	101						

Binary addition and multiplication

Addition

+	0	1
0	0	1
1	1	10

Multiplication

×	0	1
0	0	0
0	1	1

Conversion: Decimal to binary

2)166	Remainder	0
2)83	Remainder	1
2)41	Remainder	1
2)20	Remainder	0
2)10	Remainder	0
2)5	Remainder	1
2)2	Remainder	0
2)1	Remainder	1
0		

$$166_{10} = 10100110_2$$

SOURCE: Reproduced from Robert H. Roy, *The Cultures of Management* (Baltimore: Johns Hopkins University Press, 1977), p. 327.

contentious contemporary, Charles Babbage, described an "analytical engine," never built but predictive of twentieth-century events that were to come pell-mell, in a rush that has yet to abate.

Developments at the Massachusetts Institute of Technology, Columbia and Harvard universities, and at International Business Machines and the Bell Telephone Laboratories were followed at the University of Pennsylvania by ENIAC, the acronym for Electronic Numerical Integrator and Computer, the first machine to use electron tubes (Fig. 12.5). ENIAC comprised thousands of interdependent tubes, resistors, capacitors, and switches, and used "flip-flop" circuitry consisting of pairs of tubes so arranged that when one was conducting the other was not.

Using decimal numbers, ENIAC computed artillery trajectories for the Ballistic Research Laboratory of the United States Army with spectacular success, but its operations were hampered by frequent breakdowns in one or another of its thousands of interdependent

Fig. 12.5   A view of part of ENIAC. Courtesy of the Smithsonian Institution.

Fig. 12.6   Von Neumann's construct of redundant circuitry to attain very low probabilities of error. O_1, O_2, and O_3 are identical networks, each of which outputs to both "majority organs," designated M. If all three outputs are the same, the process continues. If all three have different outputs, the process is stopped. Such a system will err only if two of the three networks make the same error simultaneously (the probability of which is very small). Micro-circuits and other techniques, by redundancies greater than three and more stringent majority rules, have made possible exceedingly small probabilities of error and have provided means for identifying sources of error as well. Reproduced, by permission, from Herman H. Goldstine, *The Computer from Pascal to von Neumann*, p. 283. Copyright © 1972 by Princeton University Press.

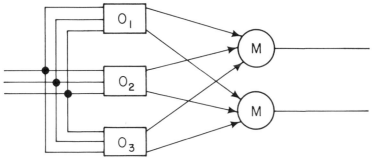

Fig. 12.7 Integrated microcircuits photographed beside the eye of a needle. Courtesy of Arthur Heigl, Director of Administrative Computing, Johns Hopkins University.

components and by insufficient memory: the capacity to store and retrieve information. ENIAC and its electron tube successors (ED-VAC, EDSAC, and IAS)[6] were also large machines, consuming power and generating heat in amounts commensurate with their size.

Operational difficulties were alleviated by use of binary numbers, redundant circuitry (Fig. 12.6), and hierarchical memory whereby information needed most quickly or most often would be stored in the machine for easiest access, that of less urgency on discs or tapes, and that of least urgency more remotely, in files or on cards.

These remedies and other innovations as well came quickly, but problems of size, power, and heat were not resolved until invention of the transistor[7] and integrated circuits on microchips (Fig. 12.7), excursions into the art and science of solid-state physics and miniaturization that have made possible the speed, accuracy, reliability, and versatility of present-day computers.

6. EDVAC (Electronic Discrete Variable Computer) was ENIAC's successor at the Moore School of the University of Pennsylvania. EDSAC (Electronic Delay Storage Automatic Computer) came from the Cavendish Laboratory of Cambridge University, (Goldstine, pp. 187 ff.), and IAS was named for the Institute for Advanced Study at Princeton, N.J. (Goldstine, chap. 2).

7. John Bardeen, Walter H. Brittain, and William Shockley of the Bell Telephone Laboratories received the Nobel Prize in 1956 for invention of the transistor.

ARCHITECTURE

Computers are composed of three principal parts (Fig. 12.8):

Primary memory
Central processing unit
Input and output units[8]

Memory

The primary memory of a computer is a repository for instructions and data. It is composed of aggregations of two-state devices, each of which can represent one binary digit, or *bit*. Each information unit in the memory is in turn composed of a fixed number of bits, each unit resides in but one location, and each location is capable of holding but one unit. Each memory unit therefore stores *contents* that are reached at an *address*.

Primary memory capacities, found to be insufficient in ENIAC, have been enormously expanded by utilization of ubiquitous, one might say innumerable, integrated circuits on microchips. These are capable of receiving, indexing, and storing instructions and data from input sources, and of retrieving and supplying information needed for processing or for output. Despite these very large increases in computer memories, capacities are often inadequate to meet ever-increasing demands.

Processing

Central processing units, characterized as the "brain of the computer,"[9] are actuated by programs provided by input sources. Program instructions and data are stored in memory, to be retrieved when, as, and if needed:

Immediately, to solve a current problem
Immediately, to record an event (and possibly its
   consequences) that may affect later attempts to solve
   problems
Periodically, as in the case of calculating and outputting a
   quarterly accounting statement

---

8. For this much oversimplified explanation of computer architecture I am indebted to Henry M. Levy and Richard H. Eckhouse, Jr., *Computer Programs and Architecture: The VAX-11* (Bedford, Mass.: Digital Press, 1980).
   9. Ibid., p. 17.

Fig. 12.8 Basic computer structure. Reproduced, by permission, from Henry M. Levy and Richard H. Eckhouse, Jr., *Computer Programming and Architecture: The VAX-11*, p. 14. Copyright © Digital Press/Digital Equipment Corp. (30 North Ave., Burlington, Mass., 01803), 1980.

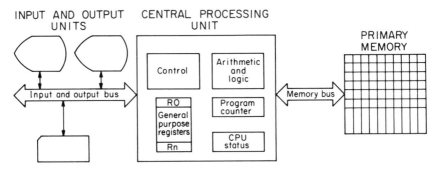

Repetitively, as in the case of directing and controlling an automated system or robot

Intermittently, as in the case of monitoring a patient's heartbeat or the course and altitude of an aircraft

Capabilities for execution go far beyond those already cited for memory storage and retrieval. Computers can perform arithmetic and mathematical operations far better than humans; they can also search, find, compare, list, sort, rank, print, display, signal, control, measure—respond to almost any functional verb one can imagine.

Input and output units

Input units provide the means by which users communicate with the machine, by providing instructions and, when necessary, data. Conversely, output units provide the means by which computers communicate with users.

The word *user* implies inputs by and outputs to humans. In the sense that the point of origin is a program prepared by a human, this is true, but in a broader operational sense, inputs can be and are provided by any kind of energy, as long as the energy increment is sufficient to initiate electrical pulses. Thus, an input can be from the keys of a console, from a magnetic tape or card deck, or from a loud noise, an irregular heartbeat, from movement of a part in a factory robot, from passage of a bar code over a laser beam in a supermarket, or from the flow of gasoline to the carburetor of an automobile engine.

Outputs can be direct, on-site responses to users who supply inputs, as, for example, by a print-out or cathode ray tube display, or they can go directly to a nonperson, such as a machine that is being controlled by the program. Outputs—and inputs as well—can be at points distant from the computer, as far away, if need be, as the moon or another planet.

PROGRAMS

All the programs previously discussed—machine-loading charts, shortest paths, articulated, balanced, continuous, and automated system designs, windmills, flyball governors, transfer machines, Monotype ribbons, and punched cards—share a common characteristic: they are all *instructions* to be obeyed by the systems and machines to which they are directed.

Computer programs also are instructions that command obedience by the machines of programmed systems. The range of possible commands is very great, obedience to properly given commands is absolute, response, except when command and machine are as far apart as Earth and Mars, is almost instantaneous, and execution is incredibly rapid and almost flawless.[10]

These advantages are overwhelmingly greater than all other preceding instructional capabilities, but they have not been easily realized.

The primary memories and central processing units of computers operate according to the two-valued binary code. Conceivably, some users might learn to write and speak in binary, but most are familiar with vocabularies and calculations that use combinations of the 26-letter alphabet and 10-digit decimal numbers. For such users, addressing instructions to a computer in long strings of zeros and ones, each arranged in a precisely correct sequence, would be not only time-consuming but conducive to error, so tedious, frustrating, and inefficient as to inhibit or prevent use of the machine.

10. "Almost instantaneous" response and "almost always flawless" execution may seem to be exaggerations. Computers sometimes seem to make spectacular mistakes, but in truth they do not. Computers do just what they are told to do, flawlessly following each program's instructions. If there are errors, they are very likely to be caused by a faulty program, not by a failure in machine execution.

To most users, responses to computer instructions seem to be instantaneous but are not, although time intervals between instructions and responses must be measured in imperceptibly small fractions of seconds—unless the instruction is to a life-seeking experiment on Mars, in which case the Earth-to-Mars command will require more than four seconds to reach its distant destination.

To resolve this translational difficulty, programming languages have been devised, each oriented toward a special purpose: FORTRAN (Formula Translator) for mathematical usage, COBOL (Common Business Oriented Language) for business, to name but two.

The vocabularies of programming languages are limited to that which is necessary for machine commands. There are no synonyms or dual parts of speech: the instruction PRINT is a verb commanding a visual output, not a noun for an art object.[11] Each vocabulary word or symbol, alone or in context, must convey but one meaning and must be used explicitly and in correct order.[12]

Programming for systems that are to be computer controlled is not a task for novices. Often, indeed almost always, there must be multiple programs for multiple purposes. Whatever the language chosen, programs must be suitable for translation, by a "compiler" or otherwise, into the binary instructions by which computers operate—and, however extensive the programs may be, they must be thoroughly tested and "debugged." Programmed system "shakedowns" can be painful.

Self-service

Some programmed systems call for self-service—that is, for direct participation by customers and clients. Most users of programmed systems are unaware of the intricacies of programming and the complex events they bring about by their patronage.

Users of telephones, for example, when they lift the transmitter from its cradle, expect to be connected with a chosen respondent, but they are not likely to give thought to the programmatic provisions that have been made for other needs and contingencies:

Busy signals
Disconnected telephones
Nonexistent numbers
Information
Time and weather service
Operator assistance

---

11. Given the command PRINT, the computer can then produce any message that can be composed with the letters, numbers, and symbols available to the machine's printer or visual display.

12. For example, the conventional multiplication sign $\times$ is not used in programming, because it too closely resembles x, X, x, X, **x**, and **X**. In FORTRAN the asterisk (*) means multiply. When accompanied by appropriate context, (*) can mean "fixed point multiply"; in a different context, it can mean "floating point multiply."

Accounting and billing for
  Toll calls
  Collect calls
  Conference calls
  Overdue accounts
Regulatory statistics
Inventory control, etc.

There are also programs that must take account of and provide for unforeseen contingencies. Programmed control of a railway classification yard, where freight cars containing cargoes of diverse weights, kinds, values, and degrees of fragility, bound for many different destinations, must be provided for in all foreseeable combinations.[13] Should there be a car with some unforeseen combinations of characteristics, that contingency may be added to the program's capability, or if the problem is not amenable to programming or is unlikely to recur, the program may call for human intervention.

PROGRAMMED SYSTEM CONTROL

So far, except by implication in Chapter 10, discussion has focused upon control over work-center sequences and the time intervals required to process units or batches of goods and services. In "perfect" automated systems, by definition devoid of human participation, there has been a tacit assumption, exemplified by the photocell and servomechanism of the color printing press, that yet another control must be exercised in both automated and programmed systems: control over the quality and accuracy of work performance.

Transfer of control over sequences and time intervals from men to machines has not been easy, as has been shown. In jobbing systems *all* controls are vested in humans. In articulated systems, control over sequences and times is transferred to system designers, but during operations quality continues to require human cognizance. So it is with balanced and continuous systems, but in these, control of quality comes to be shared between instruments and human monitors. Likewise, in automated systems, sequences and times are determined by the system's designer, while controls for these and for quality as well are exercised by machines.

In programmed systems it is the program that determines sequences, times, and quality limits.

Earlier, automated system control was said to depend upon sensing deviations from prescribed limits, responding to sensory feed-

13. *Baltimore Evening Sun*, November 11, 1968, p. C-14.

back by initiating correction, and sensing and again responding when restoration within limits has been achieved. These same restrictions apply to the control of programmed systems, but the range and acuity of sensory perception has been extended and refined, and corrective responses have been made much faster, much more delicate, and much more reliable—incredibly so, one may say.

As a consequence, the limits between which programmed operations are performed can be much smaller than hitherto has been possible. Hitting close to the center of a dime in Boston from Baltimore—or, less fancifully, targeting a space telescope on a very distant star—remains an exceedingly difficult but no longer impossible problem.[14]

Programmed operations technologies thus can respond to deviations of almost any kind, with limits made as small as circumstances may require.

VARIETY

By the means just described a computer can be programmed to initiate restorative action sensed from deviations in almost any product or service attribute that can be imagined: size, weight, position, sequence, color, odor, temperature, humidity, radioactivity—and can do this repetitively, as in other system technologies.

For articulated, balanced, and continuous systems, and, to a lesser extent, for automated systems, changes from one kind of product or service to another are difficult and costly. In programmed systems such changes are made easy and much less costly. The system can be changed to accommodate a different product or service simply by changing the program. While there is a touch of hyperbole in use of the word *simply*, it is no exaggeration to say that programming has restored to the world of operations technologies the capability for variety that once was limited to jobbing systems.

It is this capability that has closed the evolutionary circle.

Evidence that this is so is provided by the transformation of long-standing jobbing modes into systems that are now programmed. Telephony, banking, libraries, and supermarkets, once epitomized

---

14. The space telescope, when launched into orbit, must search for and focus upon some very distant star. Detection will depend upon the light energy from the star and the tolerance within which angular control must be exercised. These data, in rough approximation, are incredibly small: 10^{-23} watt for the energy difference and an angular standard deviation of 7 milliarc seconds for the tolerance. (See note 8, Chapter 10.)

Fig. 12.9 Bar code from a package of "Electrosol" purchased from a Giant Food supermarket. The identifying number of the product is printed below the bar code in arabic numerals, each digit of which is represented by the wide and narrow lines and spaces of the bar code. By scanning the configura-

tion of these lines and spaces, the laser sends requisite binary signals to the computer, and these signals respond to programs that identify and charge for the product and also perform accounting and inventory operations for use by management.

Fig 12.10 A sales receipt from Giant Food showing the item identified by the bar code in Fig. 12.9. The three items at the top of the receipt require a 5 percent sales tax, which is entered near the bottom; all other items are tax-exempt. The total, cash received, and change are listed at the bottom of the receipt, along with the date and time of purchase.

```
        GIANT FOOD - RIDGELY

          GT TONIC WTR      .50  B
          GT TONIC WTR      .49  B
          ELECTROSOL       2.59  T
          SMKD SAUSAGE     2.30  F
          SMK KIELBASA     2.49  F
          SMK KIELBASA     2.08  F
          LETTUCE           .79  F
          LESS BREAD        .95  F
          GT SHORTNING     1.15  F
          PRODUCE           .40  F
          GREEN PEPPER      .50  F
        5 LIMES            1.00  F
        2 CUCUMBERS         .66  F
          CUCUMBERS         .33  F
          CAULIFLOWER       .99  F
        2 SPRING ONION      .67  F
          TAX    .18 BAL  18.01
          CSH 20.01 CHG  2.00
   04/26/85 11:29 007437        2932
       *** ONLY THE BEST ***
```

by switchboard operators, tellers, librarians, and cashiers, are now programmed. Billions of checks, drawn on thousands of banks, in any amount, are cleared and accounted for in obedience to programmed instructions. Telephony is programmed to make whatever the connection may be that is signaled by the caller, wherever the respondent may be. For most of the items purchased in a supermarket the cashier need only pass a preprinted bar code over a laser to identify and charge for the product purchased (Figs. 12.9 and 12.10).[15] At the same time, the computer records each withdrawal from inventory and performs revenue accounting for the various classes of goods sold.

ECONOMIC AND SOCIAL CONSEQUENCES

The invention of printing in the fifteenth century had profound economic consequences. Gutenberg's jobbing system—for such it was—reduced the cost of books, made them much more available, and displaced the scribes who previously had transcribed them.[16]

Gutenberg's invention also had profound and enduring social consequences. Manuscripts that had been lettered by scribes had been available only to very limited and literate audiences. With the proliferation of books came the spread of learning that we know as the Renaissance.

The invention of textile machinery three centuries later brought about what we know as the Industrial Revolution. The economic effects of these inventions, like those of printing, were more abundant goods at lower prices, achieved by the substitution of capital for labor. The social effects were just as pervasive as those brought about by Gutenberg's technology, but they were far less salutary. Degradation, poverty, and child labor come quickly to mind when thinking of the factory system that was born of the Industrial Revolution.

15. Precursor of the various bar codes that have become ubiquitous was the "Universal Product Code" adopted by the grocery industry in 1973. Depending upon purpose, there are various modifications, but each bar code, when passed before a laser, generates binary signals that are unique for whatever the object may be (see Chapter 13). For an interesting explanation of bar codes see Aubrey Pilgrim, "Bar Code Bonanza," *PC World*, March 1985, pp. 198–207.

16. "In 1470, there were six thousand men occupied solely in transcribing manuscripts, and some years later they scarcely existed, the new process doing ten times more work than all of them put together" (D. B. Updike, *Printing Types: Their History, Forms, and Use* [Cambridge: Harvard University Press, 1937], 2:248).

These inventions and many others like them involved transfers of human skills to machines. The artistry of the calligrapher had come to be embodied in pieces of type. The dexterity of the spinster and her distaff and spindle were transferred to a spinning machine attended by a child. For the most part, these transfers were of motor, rather than mental, skills.

The invention of ENIAC and its computer and programming progeny has also involved human-to-machine transfers—not transfers of manual skills, but transfers of *information*, instructions to be carried out for purposes of computation, control, and communication.

The economic effects of these extraordinary capabilities already are evident in the ongoing transformation of earlier system technologies into systems controlled by programs. Socially, we seem headed toward increasing participation in the functioning of programmed systems by customers and clients. Computers and programs are invading every walk of life, at home, at work, and at school. It is too soon to say what these developments will be called by historians, but in the evolution of operations technologies, programmed systems certainly will be called a punctuational step.

THIRTEEN

Programmed Systems

A RESEARCH LIBRARY

At the time the investigation to be described here was undertaken, programming and computers were employed increasingly in business and industry and in military and government organizations, but not in libraries. There is reason to believe that this investigation was a pioneering effort.

About this example there is a "wheels within wheels" quality. Libraries probably are the oldest "information systems" in the world, and this case concerns application of an "information machine"— the computer—to one such system. The narrative of information about information will show not only the values to be derived from programmed control but also the difficulties of transition from old to new method—the "pangs of mutation" in the evolutionary metaphor.

PROGRAMMING IN A RESEARCH LIBRARY

Inventories abound in the living world, universally among human organizations and individuals, selectively among animals. Every inventory, even a squirrel's store of acorns, presents problems of acquisition, deposition, location, utilization, recovery, replenishment, and disposition. A library, in this case a research library,[1] is an in-

1. Beginning about 1963, with initiative from Dr. Willis C. Gore, Professor of Electrical Engineering, an exploratory study using a computer was followed by grants from the National Science Foundation to the Milton S. Eisenhower Library of the Johns Hopkins University for "An Operations Research and Systems Engineering Study of a University Library." I served as Principal Investigator of the years-long project, but credit for the pioneering events described here belongs to Dr. Benjamin

ventory for which there is a system to acquire, locate, preserve, dispense, and recover the information that comprises the library's store of knowledge.

There is, however, much more to this than meets the eye. All inventories have time horizons, most of them reasonably short. But the time horizon for a research library, the span of intended preservation, is not for a season like the squirrel's; it is forever.

Growth

That, of course, is hyperbole but it may serve to introduce a related problem: libraries are one-way inventories. They must keep information that preserves records of the past—an essential archival obligation—and, at the same time, acquire monographs, serials, maps, and documents needed to keep the facility and the research community it serves abreast with current knowledge. Some of this always growing collection cycles in and out of circulation, but there is little output to diminish the inventory's total content except by managerial decisions for removal. Such decisions about what to remove, relocate, or destroy are exceedingly difficult. They go against the grain of an organization dedicated to access and preservation.

As a consequence of this input-output imbalance, egregiously aggravated by the "knowledge explosion," growth puts severe strains upon the monetary and space resources of all libraries. Books and documents deserving of preservation and access appear in ever-increasing profusion, and serials claim continuity from volume 1, number 1.

Librarians, administrators, and trustees have learned that any new and expanded library of today will, less than two decades hence, require about twice the space and monetary resources—and double that each interval thereafter.

This kind of exponential growth has compelled attention to alternative ways of economizing while retaining convenient access to needed information. Limitation of acquisitions by local specialization and regional, national, and international sharing will be made possible by library networks dependent upon computers. It is likely that serials even more than monographs will give way to centrali-

---

F. Courtright, whose doctoral dissertation (Johns Hopkins University, 1968) has been the source of this description. Dr. Courtright has acknowledged help from many, but the lion's share of credit for conception and realization belongs to him and to Dr. Gore. So do my thanks and affection.

zation and network exchange, a development that would put severe strains upon publishers of serials and the subscribers and advertisers upon whom they depend.

Circulation control

These are changes that may lie ahead. Our concern here is with an application of programming to the control of circulation among those who use the Milton S. Eisenhower Library of the Johns Hopkins University. Other uses of programming will be cited, but the primary focus will be upon events initiated by patrons searching for, finding, borrowing, using, and returning books to be replaced upon the shelves from whence they came.

Call numbers

To be taken as given in this discourse is the call number (Fig. 13.1), the number assigned when books are cataloged following acquisi-

Fig. 13.1 Books with call numbers and borrower's ID card at time of charge-out. Reproduced from Benjamin F. Courtright, "An Operations Research Study of a University Library" (Ph.D. diss., Johns Hopkins University, 1968), p. 88.

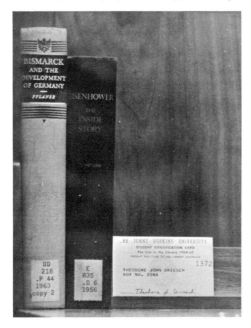

Fig. 13.2 Frequency distribution of call number lengths. *N* denotes the number of call numbers of character length *L*. The distribution is skewed toward longer call numbers. Reproduced from Courtright, p. 129.

L	N	PCT	CUM PCT
1	0	0.00000	0.0000
2	0	0.00000	0.0000
3	0	0.00000	0.0000
4	0	0.00000	0.0000
5	0	0.00000	0.0000
6	1	0.00003	0.0000
7	12	0.00041	0.0004
8	101	0.00349	0.0039
9	315	0.01088	0.0148
10	815	0.02815	0.0430
11	1438	0.04967	0.0926
12	1390	0.04801	0.1407
13	1276	0.04408	0.1847
14	2517	0.08694	0.2717
15	3777	0.13047	0.4021
16	4572	0.15793	0.5601
17	2725	0.09413	0.6542
18	2030	0.07012	0.7243
19	2351	0.08121	0.8055
20	1977	0.06829	0.8738
21	1282	0.04428	0.9181
22	827	0.02857	0.9467
23	551	0.01903	0.9657
24	349	0.01206	0.9778
25	230	0.00794	0.9857
26	175	0.00604	0.9917
27	83	0.00287	0.9946
28	55	0.00190	0.9965
29	28	0.00097	0.9975
30	26	0.00090	0.9984
31	12	0.00041	0.9988
32	12	0.00041	0.9992
33	7	0.00024	0.9994
34	9	0.00031	0.9998
35	7	0.00024	1.0000

TOTAL = 28950

tion.[2] There is but one call number for each title, and no two call numbers are alike; each uniquely identifies a title (or the title of sequential items of multivolume books and serials) in the collection.

In addition to providing identities, call numbers specify areas of scholarship (history, economics, physics, etc.), but this is a mixed blessing as new specialties (e.g., computer science, genetic engineering) and name changes in older disciplines (e.g., microbiology for bacteriology) shoulder their way upon the scene. Call numbers are, of necessity, expandable; they tend to grow long and cumbersome (Fig. 13.2), and they embody redundancy that aids human identification but, as will be shown, tends to confound the exact matching demanded by computers.

Most importantly, call numbers abound in such profusion in the library world as to make replacement by some more machine-compatible system utterly impracticable. They must, as said before, be accepted as given. However, accommodation is possible, as will be shown.

Shelving

Books in the collection are shelved in call number (alpha numeric) order, a rule that permits users to find the books for which they are looking and that specifies position when books are returned and reshelved.

This rule has the same elegant simplicity as the color-matching rule in telephony, but as collections grow, single-position-only placement creates problems. If adjacent shelves are full, a new or returned book can go into its appointed call number position only if other books are moved. As growth gradually eliminates slack space, reshelving becomes increasingly necessary. Shelf saturation in a library can lead to chaos.

Files

The circulation system of the Eisenhower Library depends upon two files: the card catalog and the shelf list.

---

2. Cataloging demands patience and skill; it is not the primary subject of this account, but it does deserve mention. Early in the project it was learned that the inventory of books awaiting cataloging contained a two-year backlog. All the books in this inventory were, in effect, in limbo; they were not represented in either card catalog or shelf list and therefore were not available for use. The backlog has since been reduced, but, more importantly, newly acquired items have been made available to borrowers by placing in the card catalog temporary entries in the form of "flimsies"that permit special arrangements for use and return to the cataloging inventory.

The card catalog is an index, in alphabetical order by author and subject (main entry), of all items in the collection. By way of illustration, note that the following cards are in the catalog for the Pflanze book shown at the left of Fig. 13.1:

One card indexed by the title, *Bismarck and the Development of Germany.* The card bears a stamp that says, "For other editions see author."

One card indexed under the author's name, Pflanze, Otto, Princeton University Press, 1963, with the notation that Copies 2 and 3 also are available, and that a fourth copy may be found in the School of Advanced International Studies in Washington.

One card indexed under Pflanze giving the SAIS call number.

One card indexed under Pflanze for the 1964 second printing. The call number differs only by the later date.

For serials such as *Chemical Abstracts,* to choose a large and important example, there may be only one call number for the series, listed by the title in a separate card catalog for serials. Because serials and certain reference books may be used only in the library and may not be charged out, we shall not be further concerned with them.

The card catalog containing approximately two million cards and organized as described, is open to all users.

Not so the shelf list. The shelf list, as it existed at the beginning of the project, contained about 300,000 cards arranged in call number order, with but one card per call number, rather than the several listed above. As such the shelf list was *the* inventory of the collection, accessible only to members of the professional staff, used constantly by them, and kept up to date by proper insertion of new cards for acquisitions and withdrawal of items disposed of.

Imputing to the shelf list representation of *all* the books in the collection must be qualified, because security has been lax for many years. There has always been an "open stacks" policy; the library is open daily until midnight, and members of the faculty (and, to a slightly lesser extent, graduate students) have almost unlimited borrowing and retention privileges. Many books no doubt have simply disappeared, and some lost books still may be rostered in the shelf list.[3]

3. The story is interesting and makes a point, but it may be apocryphal. A senior member of the faculty, on moving from his long-occupied office to another,

Despite this qualification, the shelf list was and is vital to operation of the library and, as proved to be the case, essential to the programmed system.

Staff

At the beginning of the project, members of the library staff could best be described as *professional bibliophiles*, words intended to denote and connote professionalism and love of books. They were, as most librarians seem to be, strongly imbued with a desire to serve; they were courteous and diligent in helping patrons, and knowledgeable about how to find elusive information.

They were not, however, *technologists*. The project had therefore to build a bridge, so to speak, between "two cultures."[4] Implicit was the always unacceptable premise that operations analysts, who know nothing about libraries as systems, can tell skilled and veteran librarians about a better way. There was a not always unspoken aversion to an arcane machine. And, during the transition, there were habit interferences, frustrations, and hardships, endured, not without complaint, but in good spirit withal.

THE EARLIER METHOD OF CIRCULATION CONTROL

Books to be borrowed from the library were and still are found either by browsing or by use of the card catalog for search guided by call number locations. Chosen books were then taken to the circulation desk to be charged out. The right to borrow was established by a student identification card or by faculty or staff status.

Each book in the collection had a pocket in the inside back of the case binding or cover page, in which there was a card bearing the call number and main entry and space for the signatures of successive borrowers. Opposite this, on the last left-hand page, there was affixed a form on which were rubber stamped the succession of dates by which the book should be returned and checked-in.

The librarian at the circulation desk would extract the card from the pocket, have the borrower sign for the book at the appropriate

---

returned to the library books that had been charged out for 35 years. It is still not uncommon for members of the faculty to have special collections in their private offices.

4. C. P. Snow, *The Two Cultures and the Scientific Revolution* (Cambridge: Cambridge University Press, 1959).

place on the card, and leave the card with her. She would stamp the due date on the card and beneath the last stamped date on the form at the end of the book, and the borrower would be free to depart with the book.

By repetitions of this process one signed and dated card for every book borrowed and not yet returned would accumulate at the circulation desk. These were filed in call number order. As each book was returned, its card would be removed from the file and returned to the book's pocket. The book would then be reshelved. Modest fines were required for past-due items, but this was not an across-the-board practice.

There was more to the procedure than this, but what has been said should suffice to point out serious flaws in the method.

The file of book cards in call number order was one-dimensional: it identified *what* had been borrowed, but *when* and *by whom* could be determined only by card-by-card scanning. Conceivably, a second card file indexed by due dates and perhaps a third indexed by borrowers might have been desirable, but, instead, the single file was searched but once each week for the purpose of issuing overdue and recall notices.

Security under the prevailing circumstances was necessarily lax. Undoubtedly, casual attitudes toward borrowed books, from friends as well as libraries, contributed to pilfering. Books were taken from the stacks without being charged out. There were probably also more than a few interpersonal exchanges, even for books that had been properly recorded. And given the many places a book can be (on the shelf, charged out, shelved in a wrong location, checked in but not yet reshelved, in use inside the library, out for rebinding or repair, in a faculty office, on interlibrary loan), and given the dynamics of circulation "float," any physical inventory check against the shelf list would have been impracticable, certainly counterproductive.

EXPLORATORY STEPS

In formulating objectives for the project, it came to be realized "that ultimately the contents of research libraries, viewed as information, should be processable by all relevant, adequate techniques available in the realm of information science."[5]

---

5. Courtright, p. 18.

SLOT

To progress toward this very ambitious goal, it was decided that rendering the shelf list into machine-readable form would be necessary, not so much for direct use in circulation control, but fundamental to other, related purposes important to management of the library. Accordingly, it was decided to put the shelf list on magnetic tape; hence the acronym SLOT.

This proved to be a formidable problem. The shelf list was in constant use, its cards could not be moved out of the secure area, and processing had to be done in a way that would preserve the integrity of call number order and minimize interruptions to staff use.

The first step, decided upon after consideration of many possibilities, was to reproduce ten cards per exposure on microfilm (Fig. 13.3), then return these cards to the file in call number sequence, replacing them with the next ten cards. By this means the entire contents of the 300,000-card file was quickly reduced to the size of an ordinary carton.

Possession of the mircofilm duplicate freed the project from direct dependence upon the shelf list proper but posed other problems. Microfilm is not suited to editing or marking, and catalog cards contain information needed for bibliographic but not inventory purposes. Deciding what to include led finally to these six fields of information:

Call number
Main entry (e.g., author's last name and initials)
Title
Number of pages (and volumes in a multivolume set)
Physical size in centimeters
Number of copies in the collection

These limitations and the fact that microfilm is not amenable to editing made it necessary to train transcribers to be editors while they typed. This was satisfactory for most items; for those involving questions, a signal called for editorial review.[6]

These decisions specified what should be transcribed but did not resolve the essential question of getting the transcription into machine-readable form.

At that time, circa 1965, optical scanning had risen above the data-processing horizon, but its capabilities were limited and the

6. Ibid., pp. 64, 65.

Fig. 13.3 Shelf list cards (4 cards from a 10-card-per-frame arrangement) positioned for microfilming. Reproduced from Courtright, p. 51.

P 2099
.07 A79
1961
Atkins, John William Hey, 1874–
English literary criticism; the medieval phase. Gloucester, Mass., P. Smith, 1961.
211 p. 21cm.

1. Criticism – Gt. Brit. – Hist. I. Title.

BD 161
.A 6
1962
Austin, John Langshaw, 1911-1960.
Sense and sensibilia, reconstructed from the manuscript notes by G. J. Warnock. Oxford, Clarendon Press, 1962.
144 p. 19cm.

1. Perception. 2. Knowledge, Theory of. I. Warnock, Geoffrey James, 1923–
II. Title.

PT 1492
.A2 B6
1960
Boston. Museum of Fine Arts.
Ancient Egypt as represented in the Museum of Fine Arts, Boston, by William Stevenson Smith, curator of Egyptian art. [4th ed., fully rev. Boston, 1960]
215 p. illus. 24cm.

Includes bibliography.

1. Art, Egyptian. 2. Egypt – Hist. – Ancient to 640 A. D. 3. Art - Bosto I. Smith, William Stevenson.

GN 6
.B 6
1961
Bose, Nirmal Kumar.
Cultural anthropology. [Rev.] New York, Asia Pub. House [1961]
140 p. 23cm.

Essays.

1. Ethnology. 2. Culture. I. Title.

Fig. 13.4 Part of a page from the print-out of a shelf list on tape. Reproduced from Courtright, p. 71.

```
B   67.F72              PLANCK,MK   ** WHERE IS SCIENCE GOING   ** 221P 21 2
B   67.F73V6 1961       VOGEL,F  ** ZUM PHILOSOPHISCHEN WIRKEN MAX PLANCKS   ** 255P 25 ZZ
B   67.F75              POINCARE,H ** DERNIERES PENSEES   ** 258P 18 ZZ
B   67.F8               POINCARE,H ** THE FOUNDATIONS OF SCIENCE   ** 553P 24 ZZ
B   67.P82 1958         POLANYI,M ** PERSONAL KNOWLEDGE   ** 428P 25 ZZ
B   67.P821 1959        POLANYI,M ** THE STUDY OF MAN   ** 102P 19 ZZ
B   67.P87 1959         POPPER,KR ** THE LOGIC OF SCIENTIFIC DISCOVERY   ** 479P 23 ZZ
B   67.P91 1961         PRICE,DJ ** SCIENCE SINCE BABYLON   ** 149P 22 ZZ
B   67.R18              RAMSEY,FP ** THE FOUNDATIONS OF MATHEMATICS AND OTHER LOGICAL ESSAYS   ** 292P 22 ZZ
B   67.R22              RAMSPERGER,AG ** PHILOSOPHIES OF SCIENCE   ** 304P 21 ZZ
B   67.R32 1961         READ,J ** THROUGH ALCHEMY TO CHEMISTRY   ** 206P 19 ZZ
B   67.R35              REICHENBACH,H ** FROM COPERNICUS TO EINSTEIN   ** 123P 24 ZZ
B   67.R352 1959        REICHENBACH,H ** MODERN PHILOSOPHY OF SCIENCE   ** 214P 22 ZZ
B   67.R355 1948        REICHENBACH,H ** PHILOSOPHY AND PHYSICS   ** 13P 24 ZZ
B   67.R36 1949         REICHENBACH,H ** THE THEORY OF PROBABILITY   ** 492 25 ZZ
B   67.R39 1950         RENOIRTE,F ** COSMOLOGY   ** 256P 21 ZZ
B   67.R45              REYMOND,A ** HISTORY OF THE SCIENCES IN GRECO-ROMAN ANTIQUITY   ** 245P 19 2
B   67.R5 1961          RHYS,HH ** SEVENTEENTH CENTURY SCIENCE AND ARTS   ** 137P 23 ZZ
B   67.R53 1962         RICKERT,H ** SCIENCE AND HISTORY   ** 161P 24 ZZ
B   67.R55              RILEY,IW ** FROM MYTH TO REASON   ** 327P 21 2
B   67.R57 1958         RITCHIE,AD ** STUDIES IN THE HISTORY AND METHODS OF THE SCIENCES   ** 229P 23 ZZ
B   67.R59              RITCHIE,AD ** SCIENTIFIC METHOD   ** ZZ ZZ ZZ
B   67.R7               RUEFF,J ** DES SCIENCES PHYSIQUES AUX SCIENCES MORALES   ** ZZ ZZ ZZ
B   67.R96 1946         RUSSELL,BR ** PHYSICS AND EXPERIENCE   ** 25P 18 ZZ
B   67.S2 1962          SAMBURSKY,S ** THE PHYSICAL WORLD OF LATE ANTIQUITY   ** 189P 23 ZZ
B   67.S24 1948         SARTON,G ** THE LIFE OF SCIENCE   ** 197P 22 ZZ
B   67.S25 1949         SAUER,W ** GRUNDLAGEN DER WISSENSCHAFT UND DER WISSENSCHAFTEN   ** 437P 25 ZZ
B   67.S3 1963          SCHEFFLER,I ** THE ANATOMY OF INQUIRY   ** 332V 22 ZZ
B   67.S33 1963         SCHLESINGER,G ** METHOD IN THE PHYSICAL SCIENCES   ** 140P 23 ZZ
B   67.S34 1949         SCHLICK,M ** PHILOSOPHY OF NATURE   ** 136P 23 ZZ
B   67.S36 1952         SCHRODINGER,E ** SCIENCE AND HUMANISM   ** 68P 19 ZZ
B   67.S5 1957          SHAH,I ** ORIENTAL MAGIC   ** 206P 24 ZZ
B   67.S55              SHORR,P ** SCIENCE AND SUPERSTITION IN THE EIGHTEENTH CENTURY   ** 82P 23 ZZ
B   67.S57 1959         SINGER,EA ** EXPERIENCE AND REFLECTION   ** 413P 22 ZZ
```

capital cost of equipment was too high to be feasible. A way was found, however, to transcribe cards on typewriters equipped with an optically scannable font and then to read these onto magnetic tape. With suitable guarantees of accuracy (nearly 100 percent for call numbers and less than 2 errors per 100 records for the remainder of the information), both transcription and preparation of the magnetic tapes were subcontracted to the Control Data Corporation. The total cost of this part of the operation was $14,400, and error rates were well within limits. The total cost of the entire data conversion process was $18,170 for twelve 2,400-foot reels of magnetic tape.[7] A page from the output listing, processed on the IBM 7094 and thereafter checked and revised, is shown in Fig. 13.4.

CIRCULATION CONTROL

The initial attempt to program control of circulation in the library was a pilot six-month trial in a then-existing branch library.[8] The trial was based upon a plan to substitute a punch card for the library card in the pocket of each book, with appropriate identifying information punched into each such card. Then, by providing each qualified borrower with a pocket-size identifying card punched with his or her number, each charge-out could be accomplished simply by scanning the two punch cards on automatic card-punching equipment. The book's punch card could then be returned to the book pocket and the borrower's card to the patron.

In the absence of punch cards for both books and borrowers, it was necessary to ask each patron to provide written information necessary for the preparation of punch cards for data processing, a request that generated more than a few complaints. Nevertheless, the trial sufficed to reveal the values to be derived from programming. Not only could print-outs of books in circulation be provided, in call number order, as in the shelf list, but borrowers and borrower categories could be arrayed, and so also could check-out dates, due dates, and recall notices—as could, over time, the frequency of use of each book. All this could be done without detriment to the data base, with accuracy and speed made possible by the computer.

7. Ibid., p. 67.

8. At the beginning of the project there were eleven branch libraries on the Homewood Campus at Johns Hopkins. Soon thereafter, with professional and diplomatic skill on the part of Librarian John Berthel and his staff, all branches were centralized into the Milton S. Eisenhower Library. The punch card pilot study was made by Dr. Willis C. Gore, as reported in note 1.

Fig. 13.5   Microfilming discharged books. In this view of the staff's side of the circulation desk, a librarian actuates the mechanism's spring-loaded toggle switch. The camera and mirrors are located at the lower right. Reproduced from Courtright, p. 89.

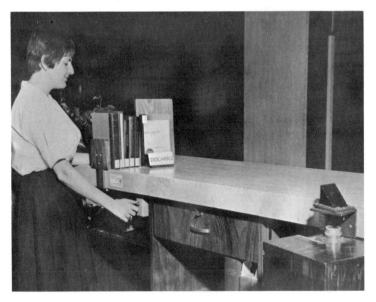

The microfilm method

As a consequence of the potential made evident by the pilot study, it was decided to implement computerized control of circulation throughout the entire library, now consolidated into one central facility on the Homewood Campus.

Two major obstacles required resolution: (1) replacing existing library cards with punch cards throughout the entire collection would be so time-consuming and expensive as to be infeasible; and (2) in the now-central library, with but a single circulation desk, peak loads would cause intolerable queues unless means could be found for a very quick charge-out procedure.

Use of a microfilm camera to photograph call numbers of books and ID cards of borrowers[9] (see Fig. 13.1) provided a satisfactory

9. Provision of identification cards had been necessary before the project was undertaken, but identification by number, as well as by name, was necessary not just for students but for all who were properly entitled to use the library. Some objected to being identified by number, considering this demeaning; a few who were

Fig. 13.6 Microfilm reader and key punch. Reproduced from Courtright, p. 107.

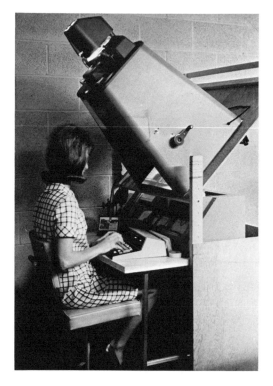

method. For each borrower, books to be charged out were placed against an easel with the borrower's card; using a spring-loaded switch (Fig. 13.5), the librarian photographed books and card, and the act of doing so automatically advanced the film to the next frame of the 100-foot reel, replaced each day by a new reel. The camera shown at the lower right of Fig. 13.5 was, except for its mirrors, located below the surface of the circulation desk, out of sight and within reach of staff only.

Borrowers returned books without being served by staff simply by depositing them on a chute leading to a collection bin. Call numbers of returned books were then photographed with a card

---

even more sensitive objected to the need to be identified. As members of the library community graduated, entered, retired, resigned, or otherwise departed or entered the scene, renewals were required, numbering close to 20,000 each year.

Fig. 13.7 A portion of the daily print-out of books in circulation. Reproduced from Courtright, p. 123.

BORROWED	DUE	CALL NUMBER	BORROWED	DUE	CALL NUMBER
2981 05/04/68	5/22	QP 475 .G7 1965	1020 04/18/68	5/22	QR 185 .A6 C5 1967
2981 04/15/68	5/15	QP 475 .P57 1967	4503 05/04/68	5/22	QR 185 .A6 S92 1965
0016 03/14/68		QP 475 .S72 1561	4624 04/11/68	5/15	QR 315 .C78
3595 05/02/68	5/22	QP 479 .G7 1963 PSYCH	0134 10/19/67		QR 360 .C8 1965
2777 05/07/68	5/22	QP 479 .15 1964	0134 10/20/67		CR 360 .S5 1965 BIOL
3595 04/23/68	5/22	CP 479 .W6 1966	0134 10/19/67		QR 360 .S6 1963 BIOL
3595 05/02/68	5/22	QP 481 .G4 1964 Q	9248 05/06/68	5/22	QR 360 .W3 1961 BIOL
2981 05/04/68	5/22	CP 481 .H453 1964	0986 04/17/68		R 15 .A55 B8 1963
3595 05/02/68	5/22	QP 481 .R8 1962	0986 04/17/68		R 15 .A55 H3 1966
15655 05/04/68	5/22	QP 514 .C32 1967	0190 05/06/68		R 15 .893 V.29 1955
0035 02/05/68		QP 521 .15 1966	0013 02/06/68		R 117 .08 W3 1918
1123 04/26/68	5/22	CP 521 .P8 1964 CHEM	0035 02/06/68		R 118 .C6 1966
0035 02/05/68		QP 528 .88 1965	0450 9/01/66		R 128 .K513 1966
4170 04/24/68	5/22	CP 551 .C44 1966	0007 04/23/68		R 131 .A1 C48 2 1967
0035 02/05/68		QP 551 .C5 1966	0007 04/23/68		R 131 .A1 J68 22 1967
0035 02/05/68		QP 551 .152 1967 V.1	1126 02/27/68	4/3	R 131 .A7 1949
0035 02/05/68		QP 551 .N45 1963 V.1 BIOL	0018 12/07/66		R 131 .G24 1929
0035 02/05/68		CP 551 .N45 1963 V.2 BIOL	2387 05/01/68	5/22	R 131 .S93 1931
0035 02/05/68		QP 551 .N45 1963 V.3 COPY3 BIOL	9188 2/23/67	3/29	R 143 .C18 1926 V.1
0035 02/05/68		CP 551 .N45 1963 V.4 COPY2 BIOL	9188 2/23/67	3/29	R 143 .C18 1926 V.2
0035 02/05/68		CP 551 .P47 1962 COPY2 BIOL	0987 04/06/68		R 152 .T85 1966
4105 05/03/68	5/22	QP 551 .P7 1965	71358 3/39/67	4/12	R 154 .859 G2
3836 05/13/68	5/22	QP 551 .S938 1966 V.4	6071 05/07/68	5/22	R 154 .8597 L85 1964 Q
0035 02/05/68		QP 551 .W65 1963	8732 7/18/67	8/2	R 159 .B2 A2 1871
2937 04/17/68	5/15	CP 601 .F4 1964	2387 05/01/58	5/22	R 487 .86 1963
0828 12/03/67		CP 601 .V5 S4 1967 V.1	4501 02/02/68	3/6	R 489 .C54 A4
0562 05/06/68		QP 601 .W5 1966	2387 05/01/68	5/22	R 489 .H21 J6 1964
0053 03/29/68		QP 701 .J4 1965 V.2 A	0943 10/14/67		R 489 .H3 K4 1966
0053 03/29/68		CP 701 .J4 1965 V.2 B	73464 6/26/67	7/19	R 489 .L6 D4 1963
0007 04/15/68		CP 751 .C4 1 1966-67	0021 4/24/67		R 489 .07 C8 1940
4709 03/28/68	5/1	QP 751 .15 1962 BIOL	0016 01/29/68		R 723 .D77 1965
4105 05/06/68	5/22	QP 801 .A5 S9 1962 BIOL	0190 04/23/68		R 737 .B79 1966
0244 8/02/67		QP 801 .H7 15 1962 V.1	6856 04/29/68	5/22	R 856 .S43 1967
0244 8/02/67		CP 801 .H7 15 1962 V.2	0007 04/10/68		R 895 .A1 A4 6 1967
0035 02/05/68		QP 801 .N8 M5 1963	6856 04/29/68	5/22	R 895 .Y3 1965
0035 03/25/68		QP 801 .P37 H6 1963	11652 04/16/68	5/15	RA 410 .A1 C6 1962
0035 02/05/68		CP 801 .P45 A5 1964	1700 05/13/68	5/22	RA 410 .M5 1960
0244 9/20/67		QP 801 .S6 D6 1965	1700 05/13/68	5/22	RA 410 .S6 1961
0828 01/22/68		QP 801 .V5 S4 1967 V.6	0035 03/05/68		RA 418 .F7 1963
0236 11/02/66		QP 903 .E9 1963	0035 02/01/68		RA 790 .W5 1968
0007 01/30/68		CR 1 .F72 27 NOS.1-4 1966	9832 04/15/68	5/15	RA 962 .F7 1963
0007 01/30/68		CR 1 .F72 27 NOS.5-12 1966	0028 6/22/67		RA 963 .J66 1875
0983 1/24/66		QR 64 .5 .15 1962	1651 04/28/68	5/22	RA 975 .S9 1967
9741 05/07/68	5/22	QR 65 .C6 1964	0986 04/28/68		RA 981 .A2 C58 1967
0035 04/15/68		QR 73 .H35 1964 BIOL	0035 01/26/68		RB 113 .G8 1966 COPY 2
2031 04/28/68	5/22	QR 73 .H35 1964 COPY 2 BIOL	0035 03/07/68		RC 180.1 .D3 1963
3663 04/03/68	5/1	QR 73 .H35 1965	0244 8/17/67		RC 261 .U46 1964
0983 9/27/66		QR 118 .A4 1962	2041 04/30/68	5/22	RC 327 .15 1961
4467 05/08/68	5/22	QR 180 .M6 V.1	0007 03/26/68		RC 367 .A35 26 1965
4467 05/08/68	5/22	CR 181 .B6 1966	0007 03/26/68		RC 367 .A35 27 1966

lettered DISCHARGE (see Fig. 13.5), after which returned books were reshelved.

Each day the developed microfilm was processed by use of a microfilm reader and key-punch machine (Fig. 13.6). The operator punched one card for each call number, including on each card the borrower's ID number, the date borrowed, date due, and complete call number. These cards were then combined with cards for other books charged out from this reel of film and all of them were read into the magnetic tape of books in circulation in call number order. From the tape a new list of books in circulation was produced for use in the library (Fig. 13.7).

To keep the print-out current it was, of course, necessary to eliminate books that had been returned and photographed as having been discharged. These microfilm images also were keypunched and then compared by the computer with call numbers already on the

Fig. 13.8   A pressure-sensitive label showing due date. Reproduced from Courtright, p. 96.

existing list. Call number pairs that matched were deleted from the tape.

Security of the collection was enhanced by centralization of the library, provision of but one point of entry to and egress from the stack areas, and surveillance by a guard of ID cards and books checked out. Inspection was made possible by affixing to each book a pressure-sensitive lable bearing the due date (Fig. 13.8). It was required that each book (and the contents of various carrying receptacles) be displayed to the guard at the point of departure.

Under the programmed check-out system, due-date labels are affixed at the circulation desk by use of a "tape shooter." Label backgrounds are color coded to provide a variety of useful signals. Due dates are automatically stamped on these labels, and the adhesive used is very resistant to removal. When books are checked in, all lables are stamped across the face with a large black X; subsequent check-outs of a title simply call for affixing a new label over any old ones already there.

Pilfering from the library may still be possible, but much less so than ever before.

Problems of implementation

For almost all the books borrowed and returned the programmed system worked well and the circulation list proved to be of great value. There were, however, some implementation difficulties that were vexing to the staff and challenging to the project team.

Call numbers and call number labels were one cause of problems. Most awkward were the indispensable, taken-as-given call numbers themselves. Many were long, as shown in Figs. 13.2 and 13.7, and all required character-by-character keying, very different from the touch typing of English text.

Every error, of whatever kind, in spacing, punctuation, transposition, or incorrect letters or figures, made during key punching of either charge-outs or check-ins would cause a mismatch, so that a returned book would not cancel the book charged out. A "nondischarged" book, as such items came to be called, would continue to appear on the circulation list, although the book itself might already be back on the shelf.

The problem was brought under control by personnel selection and training, by introduction of verification key punching, and by modifying the computer program to print out all "nondischarged" items. This list was scanned each day, and call number redundancy—for humans, not computers—made cancellations possible in most cases. Some, however, required more tedious search.

Some call number labels on old or frequently charged-out books had become so soiled that contrast between background and print, while decipherable by humans, was obscure to the camera and unreadable on microfilm. Other call number labels were wrapped around the curved spine of thin books; these were readable by humans, but the camera could not "see" around the curve.

The library already had means to duplicate soiled and curved call number labels, by replacing those too dark for microfilming and by putting duplicate labels at the top left-hand corner on the front of thin books. When photographed, such books would be placed with front rather than spine facing the lens.

Advantages

Occasional problems remained, but advantages came to outweigh disadvantages by a wide margin.

Borrowers not finding a wanted book could ascertain if it had been charged out by another patron. Unless he or she happened to know the ID number of the prior borrower, a person-to-person exchange—very undesirable in terms of control of circulation—could not be made, but the circulation desk could send a recall notice and inform the prospective user when the book had been returned and reserved for pickup.

Fig. 13.9 An overdue book notice. Reproduced from Courtright, p. 125.

OVERDUE BOOK NOTICE
THE MILTON S. EISENHOWER LIBRARY

According to our records, the books listed below were due on   FEB. 1, 1967

If not returned by   FEB. 8   an overdue fine of $5.00 per book will be assessed.
If the books are withheld an additional month, the borrower will be charged the $5.00 fine for each book, plus the replacement cost. Charges will be payable at The Financial Office, Whitehead Hall.

RETURN BOOKS TO MAIN DESK WITH THIS NOTICE

B 178.1 .L32 COPY2

E 178.1 .L35

E 178.1 .M78 1910

AYRES THOMAS RUSSELL
P.O. BOX   13

E 178.1 .T47

E 178.5 .P13

1231   THIS IS THE ONLY NOTICE THAT WILL BE SENT

Overdue notices, tedious to track and prepare under the old system, are now programmed and printed out as shown in Fig. 13.9. Fines are levied for failure to comply, and there are sanctions, some of them severe, for nonpayment.[10]

Cumulative records of books in circulation make possible data compilation as to the frequency of use of each book and, of equal or perhaps greater importance, the rostering of books that have *not* been charged out at all. This can now be done by comparing the circulation list records of frequency with the shelf list on tape. Such evidence is not sufficient cause for removal to less accessible locations or dispositions (many items have "eternal" archival value), but use frequencies and their nonuse opposites can provide information for better and wiser decisions.

Given shelf geometry, book dimensions, and SLOT, kept up to date with acquisitions and dispositions, programming can produce mapping plans for reshelving as the character and use of various parts of the collection change.

Most significantly, use of the Eisenhower Library has more than doubled since the facility itself and the programmed system began operation. This very large increase in utilization could not have been managed by the earlier method.

___

10. Among the sanctions is one of very strong persuasion: a student cannot receive his or her degree until a delinquent fine has been paid.

SUBSEQUENT DEVELOPMENTS

Those who have read the case histories of glacé fruit and telephone cable in Chapter 9 will recognize that the programmed system described here was a batch process.

Each day a "batch" of microfilm was photographed and developed and that batch of charge-outs and check-ins was keypunched and verified. The resulting batch of punch cards then went to the computer for sorting and preparation of a new tape from which the circulation list was printed and returned to the library.

When this new circulation list was put to use in the library, it did not represent the state of the system for that day. Instead, it portrayed the state of the system for the day on which the batch of film had been exposed, the interval from exposure of frame 1 to that of the last frame before changing to a new reel.

The flow process chart symbolically represented in Table 13.1 shows this sequence of processes, moves, and delays. At best, each new circulation list would become available to the library two days late, with a more likely lag of three days. In either case, all the transactions occurring during either delay, an approximate average

Table 13.1 Flow process chart of the batch circulation system

P_1^b	Microfilm books to be charged-out or discharged.
D_1^b	Await move of exposed reel to developing.
M_2^b	Move exposed reel of film to developing.
$_2D^b$	Await developing.
P_2^b	Develop, fix, dry, and rewind film.
D_2^b	Await move of film to key punch and microfilm viewer.
M_3^b	Move film to key punch.
$_3D^b$	Await keypunching.
P_3^b	Keypunch each item in each exposed frame.
$_4D^b$	Await verifying.
P_4^b	Verify keypunching, correct errors.
D_4^b	Await move of punch cards to computer.
M_5^b	Move punch cards to computer.
$_5D^b$	Await sorting and data processing.
P_5^b	Sort and process data; print new circulation list.
D_5	Await move of new list to library.
M_6	Move new circulation list to library.
P_6	Replace previous list with new list.

NOTE: The superscript b denotes items in a batch.

of 550 per day, would be akin to an in-process inventory, the where-abouts of which would be completely unknown. Any book in this in-process situation could, in effect, be anywhere; there would be no record of it until the data-processing sequence of Table 13.1 had been completed—and then another batch of in-process transactions would become temporarily "lost" in the same way.

Self-checking code numbers

To reduce this time lag and at the same time reduce the cost, tedium, and errors of call number data processing, an ingenious idea was proposed.

Each book that circulated from the collection could be uniquely represented by a six-digit number, to which could be added a seventh digit that would serve to check the accuracy and order of the pre-ceding six digits.

To compute and print out such numbers, it was proposed that an algorithm be used in which the weighted sum, in the order of 7 to 2 from left to right, of the first six digits was followed by a number which, when added to the weighted sum, was a whole-number mul-tiple of 11 (i.e., "modulo" 11). Thus, for the six-digit number 004147 one would obtain

$$7 \times 0 + 6 \times 0 + 5 \times 4 + 4 \times 1 + 3 \times 4 + 2 \times 7$$

$$= 0 + 0 + 20 + 4 + 12 + 14 = 50.$$

The next multiple of 11 being 55, the seventh digit would become 5, making the seven-digit code number 0041475. Any change in number or sequence would fail to satisfy the check digit and signal that an error had been made. Similar check digits are used exten-sively on bar codes and credit cards, generated by a variety of al-gorithms.

For the proposed seven-digit sequence it would be possible to program and print out these numbers on the IBM 1401 computer, enter them on punch cards, and print them on labels by use of the IBM 403 machine, as shown by the test run of Fig. 13.10. In actual use these numbers would be printed in a font that could be optically scanned.

Planned use

Assuming that these numbers could have been printed in sequence on dispensing tape (like that used for due-date labels), the plan was to affix whatever the next number might be in a position adjacent

Fig. 13.10   Code number labels from a test run of the IBM 403 machine.
Reproduced from Courtright, p. 138.

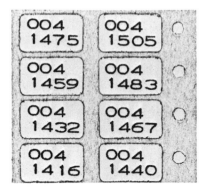

to the call number, so that both would be microfilmed in the same
frame. Both of these numbers would then be processed as before,
but henceforth, at discharge and for *all* subsequent charge-outs of
that book, only the code number, with its already established link
to the call number, would be used.

After a comparatively short time, the number of first-time charge-
outs would decrease and the number of coded transactions would
increase, and it soon would have been possible to optically scan code
numbers and make keypunching an infrequent necessity.

At the time these ideas were broached the microfilm method
was working well, but there was some resistance to yet another
technological change. More importantly, the investigator had accepted a faculty appointment elsewhere,[11] so that implementation
of code number labels would have lacked the inspiration provided
by their creator.

Bar codes

The hiatus may have been fortunate.

When the code number plan was broached, optical scanning, as
said previously, was newly upon the scene. Since then lasers and
bar codes[12] have proliferated for just the purposes needed. Their use
in circulation control may be stated briefly.

11. In 1968 Dr. Courtright was appointed to the faculty of the University of
Maryland, where he later became Professor and Chairman of the Department of
Information Science.
12. For this description of the circulation control system currently in use I am

Fig. 13.11 Borrower's ID card showing bar code and color-coded eligibility sticker.

Numbered identification cards for borrowers, one of which was shown in Fig. 13.1, have been replaced by ID cards bearing bar codes, as shown in Fig. 13.11. Each is unique for the individual identified, and some can be renewed each year as indicated by the small color-coded tag showing year of eligibility.

For use on books to be circulated, bar codes are purchased in duplicate, with "zebra stripes" to be affixed on the inside back cover of each book and, when necessary, with numbered portions affixed to xerox prints of title pages, as described below. As in the case of bar codes for ID cards, no two circulation codes are alike and each carries a check digit.

When a borrower appears at the circulation desk with one or more books to be charged out, the patron's ID card is first scanned by a fixed-head laser that reads the zebra stripe and, delving into the computer's memory, shows upon the screen of a cathode ray tube (CRT) the borrower's ID number and the appropriate due date. If there are any blocks on borrowing—for example, if the patron is not registered or his address is incorrect—a message appears on the CRT, whereupon a staff member accesses the patron's record for more details. The patron's record includes name, address, social security number, loan code indicator, patron status (e.g., normal, delinquent, etc.), and the last patron activity date.

---

obliged to Mr. David Miller, Circulation Manager, and Ms. Marilyn Petroff, Circulation Assistant, of the Eisenhower Library.

This almost instantaneous recall serves to link the borrower to the books to be checked out. These fall into two categories: books with and without a bar code.

If a book does not yet carry a bar code, a xerox print is made of the title page, to which the bar code number is affixed. The call number of the book is then written on the title page by the librarian. From this print, data are processed as described before, the purpose being to link the call number, main entry, and bar code for all future transactions. The zebra stripe portion of the bar code is then affixed to the inside back cover of the book and passed before the laser, to record the fact that this book has been charged to the borrower whose ID card has already been scanned.

If a book to be charged out has a bar code, the procedure is as described in the last paragraph: the bar code is scanned and the book is charged out to the borrower.

All borrowed books are returned by simply placing them in a container at the circulation desk. Returns are recorded as previously described, by passing bar codes before the laser with the instruction to cancel the matching charge-out records.

By these means the daily circulation list has been eliminated. If a borrower cannot find what he seeks, he may inquire at the circulation desk; the call number of the book wanted is then key-punched and the CRT screen reveals the bar code and call number, the main entry, the identity of the present borrower, his patron category, and the due date. To prevent interperson exchanges, the

Table 13.2 Laser–bar code Charge-out processes

First-time items

P_1^l	Scan borrower's bar code.
P_2^l	Xerox title page.
P_3^l	Affix bar code number to xerox print of title page.
P_4^l	Write call number on xerox print.
P_5^l	Affix bar code to inside back cover of book.
P_6^l	Scan bar code on inside back cover.
P_7^l	Affix due-date label

Repeat items

P_1^l	Scan borrower's bar code.
P_2^l	Scan bar code on inside back cover of book.
P_3^l	Affix due-date label.

NOTE: The superscript l indicates the laser method, for which there are no inter-work-center moves or delays.

identity of the person to whom the book has been charged out is not disclosed, but at the would-be borrower's request a recall may be arranged. If the wanted item has been returned, the CRT indicates that it is in the library, either reshelved or awaiting that operation.

This technology, using bar codes and laser, is another long stride ahead. The batches charted in Table 13.1 have given way to a seven-step process for each first-time charge-out and a three-step sequence for books (increasingly a majority of all charge-outs) already having bar codes. In neither case are there any inter-work-center moves or delays (Table 13.2).

The tasks of data processing first-time title pages and scanning bar codes on checked-in books are still done in batches, but in a practical sense all check-outs are matters of record as quickly as each loan is made. There is no two- or three-day in-process inventory "lost" in the system. The system has become much more effective as well as much more efficient.

This account of the impact of the information machine upon the oldest of information systems has described but one of the ways that technology has affected the operation of a research library. Much more than this has happened and much more will happen in days to come. Books, "of the printing of which there has been no end," will survive, but the ways in which books are used are certain to change.

FOURTEEN

Programmed Systems

A MONEY MARKET FUND

During the past several decades funds of various kinds have become important modes of investment.[1] By purchasing shares in a "mutual fund," investors, however large or small their means, have placed their money in the hands of professionals knowledgeable about investments and, at the same time, have secured a diversity of ownership in the institutions in which funds were invested.

There are two implications in this statement: (1) Possessing expert knowledge, the professional money manager can do better than an amateur.[2] (2) For the fee charged for management services (usually a small percentage of a fund's assets), the professional assumes fiduciary responsibility for the funds entrusted to him, accepting the obligation to place his clients' welfare foremost.

The success of mutual funds has led to their own diversification. There have come to be growth funds for those who prefer capital gain over yield, yield funds for those preferring income, bond funds, tax-exempt funds, and funds with potential for larger gain accom-

1. In 1940, Congress passed the Investment Company Act at the instigation of the Securities and Exchange Commission. The purpose of the act was to regulate "closed-end" investment trusts, some 700 of which had collapsed in the crash of 1929. (A "closed-end" trust is one whose members are limited to a chosen group, in contrast to an "open-end" fund, which is open to any who desire to purchase shares.) The act had the unexpected effect of nurturing the development of mutual funds. See Louis Engel, in collaboration with Peter Wycoff, *How to Buy Stocks*, 6th ed. (New York: Bantam Books, 1977), chap. 29.

2. Studies comparing the performance of mutual funds with various investment indices suggest that mutual funds fare no better than market averages (ibid., pp. 280ff.). To one who is not expert in the field of finance, this does not seem surprising: the amounts invested in mutual funds are so large (see note 3) as to perhaps be dominant in any market indexing method. Nor do such studies altogether negate the implication stated.

panied by higher risk. Among these and other fund categories are "money market funds," one of which is the subject of this chapter.[3]

In terms of numbers of shareholders and dollars invested, few, if any, of these kinds of funds could operate without computers and the programs by which they are instructed. Like present-day automated manufacturing, telephony, banking, libraries, and supermarkets, money management funds have been made possible by contemporary technology. The following description is intended to emphasize this point.

THE T. ROWE PRICE PRIME RESERVE FUND

The Prime Reserve Fund was established in January 1976.[4] By the end of that year 590 shareholders had invested more than $50 million in the Fund. By the end of November 1984 there were approximately 300,000 shareholders and the Fund's assets had increased to $3.2 billion (Table 14.1).

Establishment of the Prime Reserve Fund was announced in its *Prospectus*, which "sets forth concisely the information that a prospective investor should know" and by filing with the United States Securities and Exchange Commission a *Statement of Additional Information*. Both are public documents, the second being available from the company upon written request.[5]

The Prime Reserve Fund and others like it are attractive to investors because they have stable per-share values (in this case $1.00) and usually pay higher dividends than checking and passbook savings accounts. By providing for share redemptions, they are comparably liquid (Fig. 14.1).

---

3. The March 1985 issue of *Trends in Mutual Fund Activity*, published by the Research Department of the Investment Company Institute, reports the following data for three kinds of mutual funds:

	Number of funds	Total net assets billions of dollars
Open-end companies	852	156,987.8 billion
Limited maturity municipal bonds	99	33,837.5
Money market funds	332	206,707.8
	1,283	398,816.1

4. For information about the T. Rowe Price Prime Reserve Fund, Inc., I am very much indebted to Mr. George Goodman, Vice-President, T. Rowe Price Associates, Inc., and to Mr. Donald L. Fink, Executive Vice-President, DP/Associates, Inc. Both organizations have headquarters in Baltimore, Md.

5. T. Rowe Price Prime Reserve Fund, Inc., *Prospectus*, July 1, 1984, opening page.

Table 14.1 History of the T. Rowe Price Prime Reserve Fund

Date	Number of shareholders	Number of shares outstanding	On a per-share basis	
			Dividends	Net asset value
1/26/76	Original offering price			$1.00
12/31/76	590	50,855,960	$0.052	1.00
1977	1,187	33,931,260	0.050	1.00
1978	8,338	150,768,610	0.073	1.00
1979	57,343	716,575,570	0.106	1.00
1980	114,607	1,394,496,090	0.126	1.00
1981	280,012	3,183,428,356	0.161	1.00
1982	329,747	3,221,217,401	0.122	1.00
2/28/83[a]	314,100	2,831,038,533	0.014	1.00
2/29/84	288,944	2,711,186,613	0.088	1.00
2/28/85	295,714	3,185,916,455	0.099	1.00

SOURCE: Excerpted from T. Rowe Price Associates, Inc., *Annual Report*, February 28, 1985, p. 7. Per-share figures and shares outstanding reflect the 10-for-1 stock split of record that occurred on May 1, 1981.

[a]End of fiscal year changed from December 31 to February 28.

Fig. 14.1 Average annualized 7-day yields. The solid line represents the Prime Reserve Fund, the dash line the average of all taxable money funds, and the dotted line the average of all bank money market deposit accounts. Redrawn from T. Rowe Price Associates, Inc., *Third Quarter Report*, November 30, 1984, opening page.

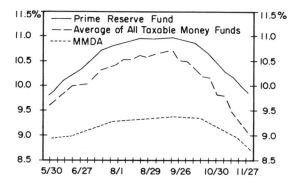

Table 14.2   Maturity diversification

Range (in days)	Percentage of net assets in each maturity range	Cumulative percentage of net assets through each maturity range	Weighted average maturity (in days)
0–30	31	31	
31–60	34	65	45
61–90	33	98	
90–120	2	100	

SOURCE: Excerpted from T. Rowe Price Associates, Inc., *Annual Report*, February 28, 1985, p. 6.

To offer these advantages, the Fund invests only in "securities which mature in one year or less," and stipulates that its "dollar-weighted average portfolio's maturity will not be over 120 days" (Table 14.2).

Consistent with this,

The Fund invests in a diversified portfolio of money market securities, limited to those described below, which are rated within the two highest categories assigned by established rating services. . . . Such securities include:

(i) U.S. Government Obligations,
(ii) U.S. Government Agency Securities,
(iii) Bank Obligations,
(iv) Commercial Paper,
(v) Short-Term Corporate Debt Securities,
(vi) Canadian Government Securities (limited to 10% of the Fund's assets),
(vii) Repurchase Agreements involving these securities,
(viii) Savings and Loan Obligations,
(ix) Foreign Securities (U.S. dollar denominated money market securities issued by foreign issuers and foreign branches of U.S. banks).

At least 80% of the Fund's assets will at all times be maintained in prime money market instruments . . . within the highest credit category assigned by established rating agencies.[6]

There are also stated precautions about foreign investments and risks associated with possible political and economic dislocations. An example of the Fund's diversification is shown in Table 14.3.

Limits are set on the amounts permitted for investment in each

6. Quoted, abridged, and rearranged from ibid., p. 1.

Table 14.3 Sector diversification

Sector	Percentage of net assets
Eurodollar negotiable CDs	43
Banking	21
Finance and credit	11
Bankers' acceptances	8
Foreign (other than Canadian)	5
Petroleum	5
Domestic negotiable CDs	4
Industrial	1
Gas and gas transmission	1
Other commercial paper	1
Total	100

SOURCE: T. Rowe Price Associates, Inc., *Annual Report*, February 28, 1985, p. 7.

diversification category, and these are expressed as percentages of the Fund's total assets. Each such percentage serves as a ceiling, and the sum of these percentages exceeds 100 percent. Each of these ratios serves as a constraint. The objective of the Fund's managers is to maximize Fund income, subject to the general constraint of not exceeding *any* allowed categorical percentage.[7]

About these limitations several observations are germane:

They were not imposed upon the Fund by government. They were conceived and stated by the Fund.

Once announced to investors, these self-imposed constraints became obligatory.

If one or more constraints were violated in such a way as to cause loss to investors, the Fund Corporation and its managers could become defendants in suits for damages.

Because constraints are expressed as percentages, they are, as said, ratios in all of which total assets are the denominator. Since share redemptions are available to all Fund investors, a single very large withdrawal could reduce the denominator and thereby increase category percentages, perhaps raising one or more above the stated limit. Such a hypothetical event

7. These percentage ceilings are not given in the *Prospectus*. They appear in the Fund's *Statement of Additional Information*.

would lie beyond management control, but the possibility, though remote, provides a cautionary motivation to maintain some slack in each category.

These abridged details are intended to introduce the programmed systems by which the Fund operates. It seems important to add that the Fund's managers and the system by which the Fund operates are obedient to the spirit as well as the letter of statutory and administrative laws governing fiduciaries, and to what may be called the analogous "administrative law," self-imposed by the Fund's *Prospectus* and *Statement of Additional Information*.

COMPUTER SYSTEMS

Management of the Fund is served by four computer systems, all of which are interconnected. One of these is in-house, in the offices of the Fund itself. Another, a service organization, maintains records of every one of the many thousands of shareholders. Another, also a service organization, accounts for the investment portfolio. And a fourth provides data required for control. Each of these services will be discussed separately.[8]

In combination, these four data banks and programs provide design redundancy to guard against system stoppages, just as circuit redundancies in computers ensure low probabilities of operational errors.

SHAREHOLDERS

Information about every shareholder must be kept current. This service, performed by DST Systems, Inc., located in Kansas City, includes the name and address of each shareholder; dates and amounts of acquisitions (the minimum amount of initial investment is $1,000), additions (the minimum for each subsequent investment is $100), and redemptions; numbers of shares held, acquired, and redeemed; dividends accrued, paid, or reinvested; annual yield for tax reporting—all of which are attributes of a multi-item accounting service demanding meticulous accuracy.

---

8. Growth in the number of organizations providing services of the kinds described has been one consequence of the "computer age."

Increments and decrements

On each business day there will be purchases of shares by new investors, additional purchases by present investors, and redemptions by others. The algebraic sum of additions and redemptions, given the growth trend previously described, is likely to be positive; if so, the net increment must be included in the next day's investments, as described below. Should the net be negative, the decrement must be subtracted. Increments and decrements are calculated by the computer and reported to the Fund at the close of each business day.

Dividends

At the close of each business day the Fund calculates net investment income by subtracting fee and operating expenses from dividend and interest revenues. The remainder, when divided by the number of shares outstanding for that day, becomes the net per-share dividend. Purchase orders for new or additional shares in the Fund must become effective before noon for that day's dividend entitlement.

Shareholder dividends are accrued daily and paid monthly, taking due account, of course, of shares owned and days held. Shareholders may elect to have their dividends reinvested in additional shares.[9]

Audits

These and all other accounting procedures are subject to audits by certified public accountants.

INVESTMENTS

The Fund accounts for investments made by the Fund's manager through the use of systems provided by the State Street Bank & Trust Company of Boston. The bank is also responsible for disbursements of dividends.

Investment decisions are made by managers of the Fund with regard to the kinds of prime short-term securities categorized above. Orders to buy (or sell) are communicated to brokers, who place the necessary orders and report confirmation to the Fund, and they in turn report the terms and conditions of acquisition (or disposition)

---

9. T. Rowe Price Prime Reserve Fund, Inc., *Prospectus*, p. 7.

to the bank for use in settling with the brokers. The accounting details for each transaction are entered by the Fund into the computer record.

The instruments in which the Fund's assets are invested are short-term, as specified in the *Prospectus*, and mature on specified dates. Payments into the Fund by investments that have reached maturity occur every business day, averaging between 2 and 5 percent of the Fund's total assets, depending upon maturation strategy. These maturities, as distinguished from dividends and interest, must be recorded and reported to the Fund so that these moneys may be reinvested.

Because the Fund's investments must be in prime securities, market values, over the short terms also specified, remain relatively stable and maintain a per-share *net asset value* very close to $1.00 (see Table 14.1). There are, however, exceptions, and these must be accounted for by the Fund in order to report the market value of the portfolio at the end of each business day.[10]

Total assets are the sum of the market values of all the securities in the Fund's portfolio at the end of each business day, augmented by the increment of new shareholder purchases and diminished by the decrement of shareholder redemptions. Each day's *total assets* then become the basis for that day's fee for management of the Fund, and the multiplicand to which each percentage constraint is applied to ascertain the amount permitted for investment in each category, as described below.

Fee and operating expenses are deducted from revenues from dividends and interest in order to compute the daily dividends due to shareholders, as described below.

INVESTMENT CONTROLS

Because all Prime Reserve Fund assets are invested in short-term securities, each with a specified maturity date, each day's maturities (or return of principal) are predictable and confirmable to the computer network. The algebraic sum of these maturations, plus pur-

---

10. For the net asset value to be held at $1.00 per share, when rounded to the nearest penny, it is necessary that the quotient of total assets divided by the number of shares outstanding be kept greater than or equal to $0.995 and less than $1.005, a tolerance of less than ±0.5 percent. This degree of stability is made possible by the combination of prime investments, stable market values over short terms to maturity, and investments that are not often traded between the time of purchase and maturity.

chase of new shares and minus redemptions, provides an amount that must be reinvested by the Fund. Daily amounts range from 2 to 5 percent of *total assets,* or from about $60 to $160 million.

All investments must be made by orders placed with brokers by 12:00 noon on the following day, lest there be loss of one day's return.[11]

Investment strategies

Decisions about what to buy from funds available each day are, as already said, constrained by allowed percentages and the prime qualities of the securities to be ordered. Orders to brokers are also guided by prudent policies directed at enhancing fund yields.

If, for example, it seems probable that there will be an upward trend in interest rates—that is, if today's interest rates are lower than expectations for subsequent days—it behooves the Fund to prefer shorter, rather than longer, terms to maturity, in order later to be able to take advantage of the expected higher yield.

Contrariwise, if it is probable that interest rates will fall, if current rates are higher than those in the near future are expected to be, it will be prudent to seek longer periods to maturity.

Investment strategies such as these are analogous to policies governing merchandise inventories, for which buyers will purchase in small lots if price declines are expected, and in larger quantities if prices are expected to rise.

P.O.L.I.C.E.

The initials for "*P*ortfolio *O*n *L*ine *I*nvestment *C*ompliance *E*xamination" form the acronym—without which no programmed system would be fashionable—for the programmed system directed to the management of each day's investments in compliance with the constraints to which the Prime Reserve Fund is committed. At some risk of repetition, P.O.L.I.C.E. is directed toward the resolution of problems such as the following:

Large sums must be invested each day.

Investment decisions (buys/sells of short-term securities) must be made each morning within a brief time period.

---

11. The range of annualized seven-day yields from the Fund's investments shown in Fig. 14.1 is from slightly less than 10 percent to about 11 percent. One day's interest on $160 million at these rates is slightly more than $43,800 for 10 percent and slightly more than $48,200 at 11 percent.

Investment strategies, not prescribed by the *Prospectus*, are dynamic, subject to changes made by investment professionals who evaluate market conditions, economic factors, interest trends, etc.

Large daily turnovers of funds create opportunities for violations.

Violations of constraints could be very costly, from class-action suits and from the costs of "undoing" any transaction that had caused a violation.

Knowledge of the flow of investments into and redemptions out of the Fund, essential to the total investment pool, are not known until early in the morning of the current business day.

To provide an "audit trail" for certified public accountants, comprehensive microfiche records are required.[12]

To make each day's investments, the Fund's managers must know not only the total amount to be invested; they must also know the total assets of the Fund at the close of each business day and the percentage of the assets invested in each of the constraining categories specified in the *Prospectus* and *Statement of Additional Information*. This means that all the securities that matured during the day must be removed from the portfolio at the Fund, at the Bank, and at DP/Associates. This data processing may take place during the current day, as matured securities are redeemed; those not cleared during the day must be accounted for after the close of the day, at night, or early the next morning.

At DP/Associates every security in the portfolio is encoded as to the category in which each belongs. When the total assets at the close of the business day have been determined, the computer produces a summation that appears in print-outs and also on the screens of cathode ray tubes (CRTs). A partial listing of the entries on such a print-out is shown in Table 14.4.

Explanation

Table 14.4 includes only 12 items from a print-out that lists several times that many. Abbreviations have been spelled out for easy comprehension.

---

12. Excerpted and paraphrased from a DP/Associates *Manual* which consists of two large volumes.

Table 14.4 Selected items from a daily on-line P.O.L.I.C.E. working document

Description	Dollars in thousands		
	Available	*Current*	*Limit*
Related issuers			155,680
New companies	150,753	4,927	155,680
Canadian government	311,361		311,361
Repurchase obligations	311,361		311,361
Industry—finance and credit	210,044	568,358	778,402
Industry—banks	322,740	455,662	778,402
Industry—petroleum	522,383	256,019	778,402
Other industry	527,880	250,522	778,402
Industry—retail	656,042	122,360	778,402
Industry—foreign governments, excluding Canada	669,356	109,046	778,402
Industry—investment dealers	715,471	62,931	778,402
Industry—electric utilities	741,216	37,186	778,402

NOTE: Par value at maturity = 45,000(000); estimated total assets = 3,113,611(000).

In each case the amount *available* is equal to the percentage *limit* minus the *current* investment. For two entries (lines 3 and 4), there are no current investments, leaving the amount available equal to the percentage constraint, in this case 10 percent of total assets ($3,113,611 × 0.10 = $311,361).

For each of the industry entries (lines 5–12), the limit is the same: 25 percent of total assets.

Line 1, showing a 5 percent limit and no data for the other two columns, is of special interest. The limit applies to issuers who, though seemingly disparate, are, in fact, related operationally, fiscally, or otherwise, as in the case of subsidiaries of a conglomerate or a holding company. In today's world of takeovers, mergers, and changes in ownership and control, identifying and reporting related issuers presents a formidable problem.

In the table note, the term "Par value at maturity = 45,000(000)" represents the amount that must be reinvested on the morning of the day covered by the CRT display and print-out.

As investments are made, buyers record each order and receive from the computer an on-line CRT display of the balance remaining to be invested.[13] By these means, the Fund's managers, working

13. As previously stated, each order is also communicated to the State Street Bank for entry into its records.

against time, always have before them the latest state of the system, to guide them toward additional investments that will diminish and ultimately deplete the investment pool—without violating *any* of the constraints pledged to shareholders.

Not touched upon here, besides other inadequacies of exposition, are provisions against error, protections against illegal or erroneous entries, and preservation of data for audits by CPAs.

These, though essential, are not germane to the central purpose: to describe a system that can and does work *only* because it has been made possible by the evolution of programming technology.

There is no way to compare the efficiency of this programmed system with any other system technology, because none could perform the multitudinous recording, calculating, and reporting that this Fund and others like it require, with voluminous and meticulous data processing during every business day.

The *effectiveness* of programming per se has made possible an important new mode of investment.

FIFTEEN

Programmed Systems

ROBOTS

In Chapter 10, transfer machines were described as automated devices capable of performing a balanced sequence of operations and moves without human assistance. Robots also fit this definition, but with an important difference: automated transfer machines are actuated and controlled by "hardware"; robots use hardware too, but in the main are actuated and controlled by "software." Robots are *programmed*.

Robots can be programmed to have the same flexibility that has been ascribed to other computer programs: change of function can be achieved by change of program. Transfer machines controlled by hardware can be adapted to design and market changes only by costly retooling or replacement; robot performance can be changed by programmatic means that are not always easy but are much less expensive.

Yet another difference is germane to this discussion. Increasingly, the programming of robots is likely to involve sensory (e.g., tactile, visual, auditory) controls that involve some of the esoterica associated with artificial intelligence. The examples that follow therefore could have been placed after Chapter 16 but are given here to serve as a bridge between programmed and "intelligent" systems.

Software programming and flexibility are augmented by other advantages of robots: coordinated and coincidental movements of great delicacy and accuracy about multiple axes (Fig. 15.1),[1] speedy

For assistance in the preparation of this chapter and for its illustrations I am very much indebted to Mr. Rollie Woodcock, Corporate Communications Manager, Intelledex, Corvallis, Oreg.

1. Sensors for the control of motion in robots are the subject of Clarence W. de Silva's paper "Motion Sensors in Industrial Robots," *Mechanical Engineering*, June 1985, pp. 40–51.

Fig. 15.1 A seven-axis robot arm. Courtesy of Intelledex, Corvallis, Oreg. Reproduced, by permission, from Rollie Woodcock, "Robots Handle PC Boards in ATE Workcells," *Electronic Packaging and Production*, September 1984.

operation, strength, repetition without boredom or fatigue, and ability to perform in environments hostile to humans.

Most important among the advantages of robots is the ability to perform a variety of tasks that once required human operators, to automate that which had been performed manually or with manual assistance. The potential for reducing or eliminating direct wages in the performance of relatively simple tasks (loading and unloading machines, spray painting, welding, assembling) has provided primary motivation for the rapid increase in the use of robots in mass production (Table 15.1).[2]

2. Robert U. Ayres and Steven M. Miller, *Robotics: Applications and Social Implications* (Cambridge, Mass.: Ballinger Publishing Co., 1983), p. 42.

Table 15.1 International robot population, February 1982

Country	Number of robots
Japan	14,246
United States	4,700
USSR	3,000
West Germany	1,420
Great Britain	713
Sweden	700
France	620
Italy	353
Czechoslovakia	330
Poland	240
Norway	210
Denmark	166
Finland	116
Australia	62
The Netherlands	56
Switzerland	50
Belgium	42
Yugoslavia	10
Total	26,924

SOURCE: Reprinted, by permission, from Ayres and Miller's *Robotics: Applications and Social Implications*, Copyright 1983, Ballinger Publishing Company, p. 7.

Wage economies no doubt will remain important in decisions to invest in robots, but as programming becomes more sophisticated and technology progresses, as it always does, other reasons will attract investments in robots. Four such possibilities will be described and illustrated here:

An automated functional test workcell
Multiple robots operated in sequence
Visual controls
The use of robots in a "clean room"

AN AUTOMATED FUNCTIONAL TEST WORKCELL

The seven-axis robot shown in Fig. 15.1 is the controlling element in the workcell diagrammed in Fig. 15.2. The arrangement is used to operate automatic test equipment for printed circuit boards.[3] The

3. The exact nature of the tests made on each circuit board is not known. Presumably, since the tests are automated, they are electric, for the discovery of circuit defects.

Fig. 15.2 Schematic arrangement of test workcell at the Electronic Manufacturing Division of Xerox Corporation. Courtesy of Intelledex, Corvallis, Oreg. Reproduced by permission.

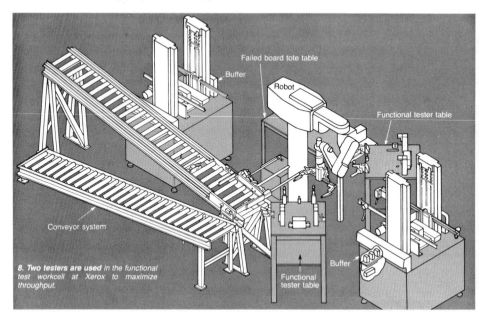

robot combines requisite dexterity and capability to adapt to different dimensions and designs in the same workcell layout, thereby providing automation *with* flexibility.

Untested circuit boards (not shown in Fig. 15.2) are delivered on the conveyor sloping downward from the left, and circuit boards found satisfactory are delivered by the robot to the conveyor sloping downward from right to left. Failed boards are delivered to the tote table.

Included in the diagram are two 20-board "buffers." These provide input and output reserves from which untested circuit boards can be taken and into which tested boards can be put during intervals of imbalance in the regular cycle.

The gripper of the robot has a two-sided paddle which, with vacuum pickup on each side, is capable of handling two circuit boards at the same time. By this means, the robot picks up the next board to be tested, moves to the test fixture, and there picks up the just-tested second board; it then rotates 180° to load the already picked up untested board into the tester and, having done so, delivers the tested board to the next vertical slot of the appropriate tote box.

The cycle then continues. The robot arm, its gripper now empty, moves to the pickup point and there finds and picks up the next untested board. (The buffer will deliver a board only upon the robot's command, which is given only when there is no board at the pickup point.) Having picked up the next untested board, the robot then goes to whichever tester has completed its task and repeats the following cycle:

Remove tested circuit board.
Rotate 180°.
Position and place untested board in tester.
Move to appropriate tote box.
Place tested board in slot.
Return to pickup point.
Pick up untested board.
Move to tester at which there is a just-tested board.

The testing of each board requires an average of 22 seconds. Annual capacity of the workcell is expected to be about 300,000 circuit boards.

The use of two testing stations enhances production by avoiding idleness. While Tester No. 1 is inspecting, the removal of a tested board from Tester No. 2 and placement there of an untested board goes on. By this time the inspection at Tester No. 1 has been completed, so without interruption the robot may go there next. Note that if the testing interval were longer, a third or even a fourth testing station might be feasible, at a cost of greater system complexity.

This much-abridged account does not do justice to the effectiveness and economy of the robot-controlled system, nor has anything been said about the delicate and discriminating qualities of the system: the dimensional accuracy with which each circuit board is placed between test pins; sensors to signal whether or not a board is properly in the gripper; the handling of occasional warped boards; capabilities for adding visual controls; and possible use of bar codes to record each circuit board's performance. There are also rules to distinguish between defective circuit boards and possibly defective testers. If a single tester rejects five boards for the same reason, that tester is checked; if there are six consecutive failures, the workcell is stopped because of probable assembly error.

Fig. 15.3   Robots assembling printed circuit boards. From component supplies shown at the left, parts are taken by each robot in turn and applied to the circuit boards in the conveyor beneath the robot arm. Movement of the boards along the conveyor is intermittent, programmed in synchrony with assembly intervals. Courtesy of Intelledex, Corvallis, Oreg. Reproduced by permission.

MULTIPLE ROBOTS IN SEQUENCE

Assembly of printed circuit boards, the operation preceding the testing just described, is made possible by arranging robots in balanced sequence, with component parts added as boards pass on a conveyor. Parts are taken by each robot from adjacent supplies and positioned and applied to the board that is before the robot. Each board remains before each robot until assembly there has been completed, at which time the conveyor moves the board to the next robot, where the assembly process goes on. When completed, boards go into the load-

Fig. 15.4 Schematic diagram of a communications network for robots arranged in sequence. The network that ties the system together allows communication back and forth between the robots in a manner that is independent of the state of the operating sequence. Courtesy of Intelledex, Corvallis, Oreg. Reproduced by permission.

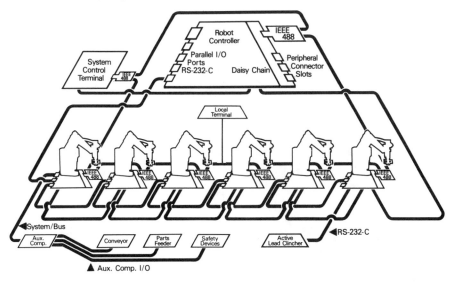

ing device at the end of the conveyor, whence they are moved to the workcell for testing. The overall assembly process is shown in Fig. 15.3.

Sequential arrangements of this kind require broad communications capability, in the manner of the network diagrammed in Fig. 15.4. Contriving and sustaining balance between and among adjacent stations by adjusting processing and moving times demands skillful design and engineering and sensitive controls, in this case through sophisticated software.

VISUAL CONTROLS

Not shown in Fig. 15.3, but possible for such use, are the cameras shown in Fig. 15.5. As seen in this illustration, the robot arm has placed a printed circuit board beneath the four cameras, which individually and collectively scan the board for the proper presence of component parts and their connectivity and polarity. The device also sorts boards according to inspection results.

Fig. 15.5 "IntelleVue" cameras used for the inspection of printed circuit board assembly. Courtesy of Intelledex, Corvallis, Oreg. Reproduced by permission.

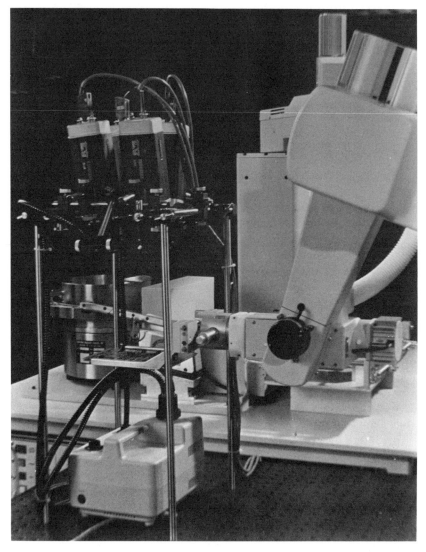

Inspection by this means requires programming for sensory controls. Once that is achieved, the system is fast and economical and immune to the errors and boredom that afflict repetitious and monotonous human scrutiny.

"CLEAN ROOM" ROBOTS

Robots, as said earlier, are capable of functioning in environments that are hostile to humans: dangerous because of possible accidents, excessively hot or cold, radioactive, submerged, or otherwise haz-

Fig. 15.6   Robot moving cassettes in a "clean room." In this operation the robot moves cassettes containing wafers into and out of a rapid thermal annealer. Triple-redundant sensing confirms that a cassette is present and properly held. Courtesy of Intelledex, Corvallis, Oreg. Reproduced by permission.

ardous or inaccessible. There is, however, another side to this coin. Robots can be advantageous in "clean room" environments: those in which humans are the source, rather than the victims, of environmental hostility.[4]

In the semiconductor industry, to which much attention already has been given, necessary pristine quality demands processing in an atmosphere almost totally devoid of particulates.[5] Humans, by moving, breathing, and perspiring, can themselves be polluters. Cumbersome protective clothing also can contribute to human discomfort and to the presence of unwanted particles.

Design requirements that permit robots to function under these demanding conditions are stringent. Sliding surfaces that can contribute particulates must be minimized, in favor of rotary joints that can have "ferrofluid seals" held in place magnetically. Exposed lubricants, used as little as possible, must have low vapor pressure. Insofar as possible, all external surfaces must be "nonshedding." Electric motors must be brushless.

Programming for process control is comparably difficult. "Robot BASIC," an extension of "Microsoft BASIC," has been devised for the delicate and repetitive operations shown in Fig. 15.6.

Once again, description has been abridged, yet is sufficient, I hope, to place robots and their increasing future uses properly in the evolutionary scheme of operations technologies.

The word *robot*, from the Czech word *robotnik*, meaning "worker or serf," first came into popular use in 1920 in the play *R. U. R.*, by Karel Čapek.[6] *R. U. R.* was first produced in Prague in 1921, later in Germany, and in 1922 simultaneously in London and New York. "The play has been translated and acted in practically every civilized country."[7]

The play, of course, is a fantasy. Čapek's robots were endowed with remarkable characteristics: language, mental and motor skills, limitless memory, boundless energy, slavish obedience—character-

4. Tom Peterson, "Anatomy of a Clean Room Robot," *Semiconductor International*, November 1984.

5. For Class 10 clean rooms, not more than 10 particles larger than 0.3μ per cubic foot of air per minute is emerging as a de facto standard.

6. Karel Čapek, *R. U. R.* (for Rossum's Universal Robots), translated from the Czech and adapted for the English stage by Nigel Playfair, edited by Harry Shefter (New York: Washington Square Press, 1973). The original English edition was published in 1923 by Oxford University Press.

7. Ibid., Readers' Supplement, p. 11.

istics that are evident in much more limited ways in the robots of today.

Like present-day robots, Čapek's at first were devoid of human emotions. Eventually, as these robots took over all the world's work, humans basked in idleness and human reproduction ceased. The robots acquired free will and seized power, but they could not reproduce themselves, so the end of both human and robotic "life" seemed imminent. At the very end of the play, however, two robots, one male and one female, discovered love and became the play's Adam and Eve.

It does not—at least not yet—seem likely that any programmed and automated robots will be male and female or, like the Tin Woodman, given a heart, but it does seem inevitable that such machines increasingly will embody, and in many ways exceed, human motor and mental capabilities.

Robots are an important part of the punctuational evolution brought about by programmed and "intelligent" technology.

SIXTEEN

"Intelligent" Systems

The words *intelligence* and *skill*, both of which defy precise definition, will be used frequently—and mayhap loosely—in the paragraphs that follow. *Skill* will describe motor strength and dexterity, and in an equally simple way *intelligence* will describe mental proficiency. Relationships between humans and the machines that humans conceive, make, and use will provide context.

Among living species, man is the quintessential maker and user of tools, and every tool ever made and used by man has embodied some element of human skill and intelligence. When Stone Age man used a rock as a tool to hurl at a quarry, his intelligence gave rise to realization that his skill would be enhanced and extended by embodiment in the missile. When his evolutionary progeny shaped and fire-hardened the point of a spear, transfers—embodiments—of mental and motor capabilities to the weapon provided even greater gain to the user.[1]

So it is with every tool that can be imagined. Each is a manifestation of a transfer process wherein human skills and intelligence are embodied in tools that are hallmarks of humankind.

For the most part, skill and intelligence are not equally evident among the innumerable tools that have been conceived, fashioned, and used. Conception and creation of the wheel and axle, like the spear, required both. Alphabets, words, and numbers were, aside from shaping and inditing, mental exercises. Domestication of the horse and invention of the windmill and plow multiplied man's motor capabilities. Assembly of Gutenberg's movable type represented the intelligence of the author and multiplied it mechanically

---

1. Kenneth P. Oakley, "Skill as a Human Possession," in *A History of Technology*, ed. Charles Singer, E. J. Holmyard, and A. R. Hall (Oxford: Oxford University Press, 1954), 1:30.

in the printing press, to produce the books that transformed the Renaissance world.

Aside from their conceptual and creative content, during the eighteenth and nineteenth centuries the machines of the Industrial Revolution were predominantly embodiments of skill transfer. Watt's steam engine provided power far beyond that previously available. The spinning jenny transferred the skill of the spinster to the ring-frame spinning machine, and that of the cottage weaver to the power loom. Craftsmanship and power were the dominant transfers and remained so for a long time.

Transfers of intelligence also were evident in Watt's flyball governor and, much earlier, in the vane that keeps a windmill fronted to the wind. But these uses of feedback did not flower, nor was the concept enunciated, until controls, analogs, and servomechanisms came to be understood and utilized in continuous and automated systems.

Now, by means of omnipresent computers controlling programmed systems, the transfer of intelligence has become transcendent, yet not complete—so far.

As "information machines" computers are embodiments of human intelligence. In some of these transfers, the performance of computers is far superior to that of humans, just as a machine's motor skills are far superior to the motor skills of the machine's human creator.

There remain, however, manifestations of skill and intelligence for which humans are at present superior to machines. Technologists devote themselves to transferring ever more skill to machines, and computer scientists are assiduous and successful in transferring more and more of what I have called "intelligence" to computers. This esoteric, arcane activity is called "artificial intelligence" (AI). It consists, in what I consider to be the best definition of AI, in seeking "to make computers do things at which, at the moment, people are better."[2]

Computer scientists have been successful in making computers do things at which humans have been better by devising new algorithms that, so to speak, imitate human abilities, by creating new programming languages that enable computers to do ever more extraordinary things, and by introducing changes in computer architecture. A few of these accomplishments and some as yet unrealized ambitions are described below.

2. Elaine Rich, *Artificial Intelligence* (New York: McGraw-Hill Book Co., 1983), p. 1.

ALGORITHMS

Heuristic Search

Given the need to search among large numbers of items for purposes of comparison—for example, reports to the Internal Revenue Service of dividends paid by corporations and received by shareholders—computers perform much better than humans; they are faster, more accurate, and not given to fatigue. There are times, however, when searching and comparing confronts a "combinatorial explosion" which is so great that complete and systematic search must give way to a more efficient method that is likely but not certain to succeed.

As a trivial but not altogether meaningless example, imagine that the problem is to find a word by unscrambling the letters *u, f, t, l, u, m, h,* and *o.* A computer could do this by alphabetically arranging the sequence in all its possible combinations (8! = 40,320), then comparing each with an alphabetic array of eight-letter English words, and printing out or displaying each match found.

By the time this systematic and comprehensive search had rostered all the matches found, it is likely that a human would have discovered *mouthful* as an answer to the problem. Success would have derived from *knowledge:* of a vocabulary learned from childhood, of *ful* as a common suffix, of *th* and *ou* as frequently occurring combinations, of *lh, hl,* and *uu* as unlikely pairs. By following *outh* with *ful, mouthful* would be quickly identified as *an* answer—not necessarily the *only* answer, but still a satisfactory solution.

Despite the triviality of the problem, it does illustrate the difference between what may be called two "thought processes": the strictly logical, totally comprehensive search by the machine for all possible answers, and the partly intuitive, partly logical "heuristic"[3] search by the human for a satisfactory answer.

The contrast between these two processes leads to a question: can computers be programmed to search more efficiently, less comprehensively (excluding that which is extraneous, unnecessary, or infeasible)—in short, to search for satisfactory but not necessarily best answers?

The answer to this question is yes.

Heuristic programming could and would be contrived to un-

---

3. "The word *heuristic* comes from the Greek word *heuriskein,* meaning 'to discover,' which is also the origin of *eureka,* derived from Archimedes' reputed exclamation *heurika* (for 'I have found'), uttered when he discovered a method for determining the purity of gold" (quoted from Rich, p. 35n). An added dictionary definition seems apropos: "valuable for empirical research but unproved or incapable of proof" (*Webster's Seventh New Collegiate Dictionary*).

scramble eight letters—any eight letters—but also to unscramble more than eight letters, a process involving factorials far larger than 40,320 (11! ≈ 40,000,000, for example), combinations far beyond the reach of most humans.

As a more realistic case in point, consider what has been called the "traveling salesman problem."[4] Suppose that a salesman is to visit 20 cities, each exactly once, with a final return to the point of departure. In making this trip the salesman should, of course, take the shortest route.

For 20 cities there are (20 − 1)! possible paths, with search time proportional to 20!. These numbers are much larger then those involved in the eight-letter scramble (19! ≈ 1.2 × 10^{17} and 20! ≈ 2.4 × 10^{18}). They are so large, in fact, that the salesman would be well advised to sally forth, reasoning that his own choice of route would be likely to cover the circuit and get him home faster and at less cost than awaiting the computer's discovery of the minimum-distance route.

Depending upon the number of cities to be visited, the pattern of their locations, and how far the salesman must travel to get home, it seems likely that the salesman's own choice of route would be better—that is, would cost less—than a comprehensive search by the computer to find *the* shortest path.

By using heuristic search, the programmer can do much better than investigate more than a quadrillion possible paths. Using a technique called "branch and bound," he can program the computer to generate complete paths and keep track of the shortest found so far. As the exploration proceeds, each new path can be abandoned as soon as the computer determines that a partially completed distance exceeds the shortest path already found. This method remains tedious but is guaranteed to find the shortest path.

By sacrificing certitude for efficiency, a "nearest neighbor algorithm" also can be programmed:

1. Arbitrarily select a starting city.
2. To select the next city, look at all the cities not yet visited. Select the one closest to the current city. Go to it next.
3. Repeat step 2 until all the cities have been visited.

This procedure executes in time proportional to N squared, a significant improvement over $N!$[5]

The route found by this means is very likely to be satisfactory but may not be the shortest path. In using this method, it should be

---

4. Excerpted from Rich, p. 34.
5. Quoted, by permission, from Rich, p. 35.

pointed out, the programmer is doing exactly what people normally do: find a satisfactory answer, not necessarily the best answer.[6]

Expert Programming

Unscrambling letters in order to recognize a word requires of humans knowledge of a vocabulary and of frequently occurring subassemblies of letters. Knowledge of this kind is possessed by many, and heuristic searches, whether by men and women in their daily lives or by computer programs, are commonplace.[7]

There are other "combinatorially explosive" problems, however, that demand knowledge that is not commonplace but esoteric. To write an "expert program," or to participate in such an endeavor, requires that there be a body of knowledge about whatever the problem may be, and that this knowledge be possessed as comprehensively as possible by whoever the expert may be.

Games of the kind that have played so important a part in the development of statistics and probability[8] have been of analogous interest to expert programmers, with chessmen the focus of interest rather than dice. Early predictions that computers could be programmed to play "perfect" games of chess have not been realized, however, there being roughly 10^{120} possible board positions.[9]

To examine each possible move and countermove is obviously out of the question. The problem seems insuperable. Nevertheless, by not examining each impending move, for which there are, on average, 35 *legal* possibilities, but instead using a *"plausible*-move generator,"[10] one can program computers to win at chess over most who play, save perhaps chess masters.

More pragmatically, expert programming has been of value in medical diagnosis (see Chapter 17), electronic design, and scientific analysis. In the exceedingly difficult domain of medical diagnosis, hands-on clinicians continue to make vital decisions as to maladies and therapies, but expert programming can provide information not

---

6. See Herbert A. Simon, *Models of Man* (New York: John Wiley & Sons, 1957), p. 204. Simon calls this "satisficing."

7. A colleague of mine who is knowledgeable about computers has declared that he "uses heuristic search all the time," but he also denies that the method is a manifestation of artificial intelligence.

8. F. N. David, *Games, Gods, and Gamblers* (New York: Hafner Publishing Co., 1962).

9. Rich, p. 26.

10. Ibid., p. 114.

only to support or confirm diagnoses but also to augment the expert program. Computers have proved invaluable, for instance, in image enhancement and interpretation.

Linear Programming

As seen in the case of the "nearest neighbor algorithm," decisions often must be made to be satisfactory rather than perfect. So it is when constraints of time, cost, prospective revenue, and conflicting values preclude searches for optima, favoring instead actions that are good but not necessarily best. Such decisions abound in the allocation of resources, where there are many interacting and conflicting variables:

How shall funds be allocated?
What products shall be produced? How many of each?
How shall they be marketed? At what prices?
What kinds, classes, and numbers of workers shall be
    employed? How shall they be paid?
What raw materials shall be procured? From what sources?
    In what quantities? At what prices?
What shall be spent for advertising? In what media?

These and many other considerations must be weighed against prospective revenue to yield acceptable profit.

About fifty years ago the method of linear programming was conceived[11] to provide better answers to multivariable, complex problems such as these. Practical use of the method, however, was dependent upon two things: development of the computer and creation of the "simplex method" by George Dantzig.[12] The conjunction of rapid improvements in computers and increased application of Dantzig's algorithm, has made linear programming a technique of great utility and frequent use for the solution of all kinds of problems involving resource allocations.

There remain, however, problems of such complexity, involving so many variables, as to require inordinate time for solution, or to

11. The creation of linear programming has been attributed by Vincent V. McRae ("Linear Programming," in *Operations Research and Systems Engineering* [Baltimore: Johns Hopkins Press, 1960], pp. 366, 367) to Professor T. C. Koopmans of Columbia University. McRae also cited the contributions of others.

12. George Dantzig, A. Orden, and P. Wolfe, "The Generalized Simplex Method for Minimizing a Linear Form under Linear Inequality Restraints" (RAND Memorandum RM-1264, 1954).

lie beyond the capacity of even the largest computers. These barriers may have been broken by a recent discovery at the Bell Telephone Laboratories, a discovery that permits feasibly rapid solutions to hitherto intractable problems involving thousands of variables.[13] If these promises are realized, as seems possible, yet another algorithm will have transferred additional human intelligence to the machine.

LANGUAGE

Some years ago I had the privilege of arranging and attending a seminar at which a friend and colleague[14] spoke of efforts to program a computer to play digital music. His early efforts were frustrated by realization. that musical notation is not like COBOL or FORTRAN: musical notation is *not* unambiguous. If, for example, a musical score calls for five equally stressed grace notes, and if the computer program plays them as written, the result is mechanical rather than artistic, much like the music of the old-fashioned player piano. Concert artists play musical scores *interpretively*, not literally.[15]

To make computer music indistinguishable from concert recordings, it was necessary for the experimenter to augment musical notation with expert interpretation. He succeeded in this, and in doing so, revealed possibilities for composing music, not just for human performance but for computers, compositions that were unrestrained by the motor and mental limitations of humans.

By succeeding in this difficult endeavor, the experimenter made it possible for computers to do something that hitherto had been done better by humans, and imparted to computers yet another portion of human skill and human intelligence. By our chosen definition, this was a contribution to artificial intelligence. It was also

13. Gina Kolata, "A Fast Way to Solve Hard Problems," *Science*, September 21, 1984, pp. 1179, 1180. Narendra Karmarkar, a mathematician at the AT&T Bell Laboratories, is credited with discovery of the method.

14. Dr. Robert P. Rich, Principal Professional Staff, Applied Physics Laboratory, and Associate Professor of Biomedical Engineering, Johns Hopkins University. Dr. Rich is the father of Dr. Elaine Rich. To both I am very much indebted.

15. Rich, chap. 9. See also "Artificial Intelligence (I): Into the World," *Science*, February 24, 1984, pp. 802–5; "The Necessity of Knowledge," ibid., March 23, 1984, pp. 1279–82; and "Natural Language Understanding," ibid., April 27, 1984, pp. 372–74.

Fig. 16.1   The Tower of Hanoi. The objective is to transfer the four discs from pin *A* at the left to pin *B* in the center, using pin *C* for transfers. Only one disc may be moved at a time, and no disc may be placed upon another of smaller diameter. Redrawn from C. L. Liu, *Introduction to Combinatorial Mathematics* (New York: McGraw-Hill Book Co., 1968), p. 67.

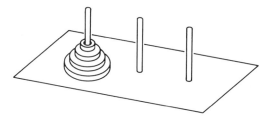

a linguistic accomplishment, for it involved devising instructions to the computer in a new and different language. Others have done this too in some of the ways described below.

LISP: The Tower of Hanoi

The Tower of Hanoi,[16] a mythical problem of long ago, involved the notion that the universe would end when monks in a Far East temple succeeded in transferring 64 discs from one of three pins to another. The myth was not as apocalyptic as it first seemed, however, because of the rules of the game: only one disc could be moved at a time; each disc was of diameter different from all the others; and no disc could be placed upon another of smaller diameter.

These rules and the 64 discs to which they were applied created a problem of great complexity. The complexity remains, but a solution does exist, by means of an algorithm by which a sequence of moves ensures success. The discs shown in Fig. 16.1 are to be moved from pin *A* at the left to pin *B* in the center, using pin *C* at the right for purposes of transfer. Four is not a large number, yet it is sufficient to illustrate method and complexity.

The algorithm states that

$$M_n(A,B,C) = M_{n-1}(A,C,B) \cdot M_1(A,B,C) \cdot M_{n-1}(C,B,A), \quad (16.1)$$

where the left side declares that *n* discs are to be moved from *A* to

16. P. H. Winston and B. K. Horn, *LISP* (New York: Addison-Wesley Publishing Co., 1981).

B, with C used for transfer. The centered dot is a sequencing operator, calling for three actions to be performed in left-to-right order.[17]

For four discs this expression becomes

$$M_4(A,B,C) = M_3(A,C,B) \cdot M_1(A,B,C) \cdot M_3(C,B,A). \tag{16.2}$$

As the right side of this equation stands, it twice calls for a rule violation, by specifying that three discs shall be moved, first from A to C and subsequently from C to B.

To surmount this difficulty we must assume that such moves ultimately may be possible with single discs if one seeks single disc moves for each M_3 expression. This is done by using the algorithm for each of the two M_3 expressions in equation (16.2):

$$M_3(A,C,B) = M_2(A,B,C) \cdot M_1(A,C,B) \cdot M_2(B,C,A). \tag{16.3}$$

$$M_3(C,B,A) = M_2(C,A,B) \cdot M_1(C,B,A) \cdot M_2(A,B,C). \tag{16.4}$$

Note that in equation (16.3) C and B have exchanged positions, as was the case on the left side as well. Similarly, A and C have exchanged positions in equation (16.4).

As equations (16.3) and (16.4) stand, each also twice calls for a rule violation. We must therefore once more apply the algorithm to all four M_2 expressions:

$$M_2(A,B,C) = M_1(A,C,B) \cdot M_1(A,B,C) \cdot M_1(C,B,A). \tag{16.5}$$

$$M_2(B,C,A) = M_1(B,A,C) \cdot M_1(B,C,A) \cdot M_1(A,C,B). \tag{16.6}$$

$$M_2(C,A,B) = M_1(C,B,A) \cdot M_1(C,A,B) \cdot M_1(B,A,C). \tag{16.7}$$

$$M_2(A,B,C) = M_1(A,C,B) \cdot M_1(A,B,C) \cdot M_1(C,B,A). \tag{16.8}$$

Having now realized single disc moves, we may assemble them in sequence, including the proper ordering of the M_1 expressions in equations (16.2), (16.3), and (16.4). This has been done in Table 16.1 in moves 4, 8, and 12, respectively, as shown in equations (16.3), (16.2), and (16.4).

The four discs moved from pin A to pin B in strict accordance with the monks' rules reveal both the value of the algorithm in prescribing a successful sequence of moves and the tedium and exactitude demanded by the game, even for so small a number of discs.

---

17. For definition of the centered dot as a sequencing operator, I am again indebted to Dr. Elaine Rich.

Table 16.1 Sequence of 15 moves required to transfer four discs from Pin *A* to Pin *B*, using Pin *C* for transfers. The letters *w*, *x*, *y*, and *z* indicate smallest, next larger, still larger, and largest disc diameter, respectively.

Move no.	Sequence of instructions		Move one disc from	Move one disc to	Pin A	Pin B	Pin C
	$M_4(A,B,C) =$				w x y z	0	0
1	$M_1(A,C,B)$	·	A	C	x y z	0	w
2	$M_1(A,B,C)$	·	A	B	y z	x	w
3	$M_1(C,B,A)$	·	C	B	y z	w x	0
4	$M_1(A,C,B)$	·	A	C	z	w x	y
5	$M_1(B,A,C)$	·	B	A	w z	x	y
6	$M_1(B,C,A)$	·	B	C	w z	0	x y
7	$M_1(A,C,B)$	·	A	C	z	0	w x y
8	$M_1(A,B,C)$	·	A	B	0	z	w x y
9	$M_1(C,B,A)$	·	C	B	0	w z	x y
10	$M_1(C,A,B)$	·	C	A	x	w z	y
11	$M_1(B,A,C)$	·	B	A	w x	z	y
12	$M_1(C,B,A)$	·	C	B	w x	y z	0
13	$M_1(A,C,B)$	·	A	C	x	y z	w
14	$M_1(A,B,C)$	·	A	B	0	x y z	w
15	$M_1(C,B,A)$	·	C	B	0	w x y z	0

Let that number increase to 5 discs, and the number of moves increases to 31; for 6 discs 63 moves are required; for 10 discs the number is 1,023; and for 20 discs more than a million moves are required. All must be carried out in the exact sequence specified by the algorithm. For n discs the number of moves is $2^n - 1$. The monks' problem therefore confronted them with $2^{64} - 1 \approx 18.45 \times 10^{18}$ moves, all to be made in correct sequence. If in the temple they are still working on the problem, they need not worry about the end of the universe!

So far, this example has depended upon the algorithm and its imaginative creator. It may also be evident that, with the algorithm, the problem can be solved for any number of discs much better by a computer than by a human. Here, too, intelligence has been contributed to the machine, not just for solving this problem, but as a linguistic simplification for this and diverse other kinds of problems. Figure 16.2 is the program for the Tower of Hanoi in the important program language LISP. In only 12 lines it provides for all the recursions required for the computer to sequence the moves for n discs. Other computer languages, FORTRAN, for example, would require much lengthier instructions. LISP, therefore, is another contribution that fits our chosen definition of artificial intelligence.[18]

SHRDLU

SHRDLU[19] is not the acronym for a computer program language, as might be supposed, but the name given to a simulated robot operating in a "blocks world" in which it moves three-dimensional shapes

---

18. Some may feel that creation of the algorithm and the LISP program for the Tower of Hanoi problem is nothing but an exercise in mathematics and computer science, and therefore of no practical value. One can never tell. A late colleague and friend once solved what he called "The Case of the Homicidal Chauffeur," in which the murderous driver of a fast and powerful car sought to kill a slower and weaker but more maneuverable pedestrian. The questions were: What course should the pedestrian take to avoid being killed? What course should the driver take to succeed in killing the pedestrian? The problem and the solutions appear to be trivial until the characters are changed to a destroyer seeking to "kill" a slower and more vulnerable submerged submarine. See Rufus Isaacs, *Differential Games* (New York: John Wiley & Sons, 1965).

19. Computer scientists are ingenious in creating new acronyms that are as useful to them as they are baffling to the uninitiated. As a one-time printer, I find Terry Winograd's SHRDLU more than ordinarily interesting because it derives from the keyboard of Ottmar Mergenthaler's Linotype. When the machine compositor made an error and wished to provide a signal to the proofreader and hand-compositor

Fig. 16.2 LISP program for the Tower of Hanoi. Reproduced, by permission, from P. H. Winston and B. K. Horn, *LISP*, p. 89. © 1981, Addison-Wesley, Reading, Mass.

```
(DEFUN TOWER-OF-HANOI (N) (TRANSFER 'A 'B 'C N))        ;N disks on A first.

(DEFUN MOVE-DISK (FROM TO)
       (LIST (LIST 'MOVE 'DISK 'FROM FROM 'TO TO)))     ;Build instruction.

(DEFUN TRANSFER (FROM TO SPARE NUMBER)
       (COND ((EQUAL NUMBER 1) (MOVE-DISK FROM TO))      ;Transfer one disk.
             (T (APPEND (TRANSFER FROM                   ;Move from FROM
                                  SPARE                  ;to SPARE
                                  TO                     ;using TO as space
                                  (SUB1 NUMBER))         ;(n-1) disks.
                        (MOVE-DISK FROM TO)              ;Move lowest disk.
                        (TRANSFER SPARE                  ;Move from SPARE
                                  TO                     ;to TO
                                  FROM                   ;using FROM as space
                                  (SUB1 NUMBER)))))))    ;(n-1) disks.
```

in response to verbal commands.[20] In the simulation there are no real blocks nor is there any physical movement of them; instead, images are located in an assigned space, identified by name, shape, and color, and displayed on the scope of a cathode ray tube. Participants in the dialogue are SHRDLU and a person called FRIEND. Communication is by typed instructions and responses.

The manner of this experiment with commands and responses in "natural language" may be illustrated by a portion of the dialogue itself, printed with FRIEND's instructions in boldface, computer responses in small caps, and the experimenter's comments in roman text:

1. **pick up a big red block.**
   OK.
   The system answers "OK" when it carries out a command. In order to pick up the red block, it had to clear it off by finding a space for the green one and moving the green one away. [Fig. 16.3]

---

to remove the incorrect slug from the galley of type, he ran his left and right index fingers down the outside rows of keys, to cast ETAOIN SHRDLU. Since hot-metal Linotype is rapidly giving way to computerized photocomposition, Winograd's choice of acronym has a certain flair.

20. Terry Winograd, *Understanding Natural Language* (New York: Academic Press, 1976).

Fig. 16.3 "Pick up the big red block." Reproduced, by permission, from Terry Winograd, *Understanding Natural Language* (New York: Academic Press, 1976), p. 8.

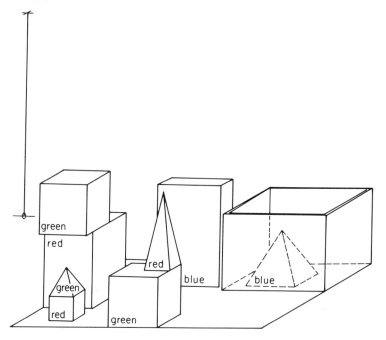

Fig. 16.4 "Find a block which is bigger than the one you are holding and put it into the box." Reproduced, by permission, from Winograd, p. 9.

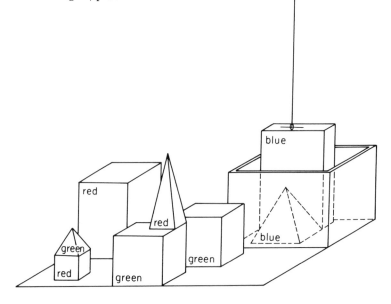

2. **grasp the pyramid.**
   I DON'T UNDERSTAND WHICH PYRAMID YOU MEAN.
   There are three pyramids in the scene, so the program does nothing.
3. **find a block which is taller than the one you are holding and put it into the box.**
   The system has to figure out what is meant by "one" and "it," as well as using deduction to decide which block it is holding.
   BY "IT," I ASSUME YOU MEAN THE BLOCK WHICH IS TALLER THAN THE ONE I AM HOLDING.
   Printed out to keep us informed.
   OK.
   The hand has to set down the red block to pick up the taller blue one. Figure [16.4] shows the results of the system's manipulations so far.
4. **what does the box contain?**
   THE BLUE PYRAMID AND THE BLUE BLOCK
5. **what is the pyramid supported by?**
   Earlier the system responded to "the pyramid" by saying, "I don't understand which pyramid you mean." Now it assumes that we mean the one just mentioned.
   THE BOX.
6. **how many blocks are not in the box?**
   Negatives such as "not" and "no" are handled whether they appear in clauses or noun groups (such as "no boxes").
   FOUR OF THEM.[21]

The dialogue continues to provide evidence of additional capabilities for several more pages, but the abridgement given here may be sufficient to show that SHRDLU has been given the ability to integrate what it knows about the "blocks world" with what it knows about English. In response to the first question, it had to find space for and move away the green block in order to pick up the red one. When given an ambiguous second instruction, it asked for clarification. In answer to question 5, it made an assumption. SHRDLU has not made the computer "like a man," but a step has been taken toward that still-distant goal.

At the heart of this transfer has been the utilization of several programming languages (e.g., LISP, PROGRAMMER, PLANNER), impartation to the computer of a vocabulary for dealing with its circumscribed "blocks world," the objects contained in that world, their spatial relations, commands applicable to them, memory, and such elusive properties as deduction, meaning, semantics, interpretation, and ambiguity. In short, SHRDLU has successfully embodied

---

21. Quoted, by permission, from ibid., pp. 8–10.

in the machine mental and motor talents once possessed exclusively by humans—an imaginative achievement in artificial intelligence.

ARCHITECTURE

Perhaps the most important means by which the capabilities of computers are likely to be enhanced is some form of "parallelism," a basic change from the "serial, one-step-at-a-time architecture laid out by computer pioneer John von Neumann in the 1940s."[22] Numerous investigators throughout the world are assiduously pursuing the various ways by which parallel circuitry can best be achieved, to the end that there may be very large interconnected systems—"supercomputers"—supported by vast amounts of software and serving whole communities of users.[23]

The "architecture" of the brain, as yet only partly understood, embodies many parallel circuits. In this sense, human anatomy, physiology, and neurology once again provide a model for the architecture of computers. For a tyro to predict the course these developments will take would be pretentious, yet one prediction germane to this discourse can be made: more, probably much more, of man's intelligence will be transferred to the machine.

CONCLUSION

The argument presented in this chapter can be stated in the manner of a syllogism:

> Among living species, man is the preeminent, almost exclusive, maker and user of tools.
>
> Every tool, of whatever kind, embodies elements of human motor and mental capabilities, defined here as skill and intelligence.
>
> The computer, a recent and revolutionary device, embodies more of man's intelligence than any predecessor tool or machine, and it is already doing many things better than can be done by humans.

22. "Artificial Intelligence in Parallel," *Science*, August 10, 1984, pp. 608–10.
23. Ibid. I am also indebted to Dr. Rodger D. Parker, Professor of Health Policy and Management and Professor of Mathematical Sciences, Johns Hopkins University, for some of the developments reported here.

Additional transfers of intelligence will permit computers to do more and more things better than can be done at the moment by humans.

In accordance with this definition, it must be concluded that artificial intelligence is here to stay—and that there will be more to come.[24]

This conclusion, which is intended to be forthright, may be accepted by some, but it certainly will not be accepted by all. Those who disagree can rightly say that each of the accomplishments reported here has been created by a human. Other, similar creations by humans have been acknowledged. But transfer to the machine of the whole gamut of human capabilities and attributes—emotion, motivation, perception, cognition, language, adaptability—is regarded by many as impossible, while others consider complete transfer a necessary condition of "real" artificial intelligence. Still others, perhaps not altogether seriously, have aspired to make a computer "like a man,"[25] a mechanistic goal that offends some and makes skeptics of others. It does not seem likely that the aspiration will be realized, yet it will be approached, if only distantly.

In the context of this book, evolution has not yet produced "intelligent systems" capable of self-correction, or self-adaptable to that which has not been provided for. But *that* goal may yet be reached.

24. Richard M. Cyert, "The Plight of Manufacturing: What Can Be Done?" *Issues in Science and Technology,* Summer 1985, pp. 87–100.

25. Charles Babbage's never-built analytical engine was intended to be a "universal machine" (Philip Morrison and Emily Morrison, eds., *Charles Babbage and His Calculating Engines* [New York: Dover Publications, 1961]). Similar attributes were accorded to the "Turing machine" and to likenesses between computers and the human brain (Andrew Hodges, *Alan Turing: The Enigma* [New York: Simon & Schuster, 1983], pp. 96ff). Notable scholars who have philosophized in this vein include W. Ross Ashby, John von Neumann, Norbert Weiner, and Herbert A. Simon.

SEVENTEEN

"Intelligent" Systems

EXPERT PROGRAMMING IN
MEDICAL DIAGNOSIS

Two branches of medical practice, diagnosis and treatment, are evident, although not always separately distinguished, in most patient-physician consultations. Based upon perception, measurement, communication of symptoms, patient history, and disease prevalence, the physician judges what he believes the patient's illness to be. That judgment, and decisions about the treatment to follow, are derived from "differential diagnosis," the symptomatic differences between the perceived cause of an illness and other illnesses excluded by the diagnostic process.

There are four possible outcomes to diagnostic decisions:

A *true-positive* diagnosis correctly identifies an illness.

A *true-negative* diagnosis correctly finds an illness to be absent.

A *false-positive* diagnosis identifies an illness from which the patient does not suffer.

A *false-negative* diagnosis finds an illness to be absent when it is actually present.

Among these possibilities, the first is best, and fortunately most diagnoses fall into this category. Illnesses are correctly diagnosed and treated, and unless an illness is fatal, patients are likely to recover.

The true-negative diagnosis is correct, but all diagnositc verification depends upon subsequent events, so there is a certain risk should a patient be sent away untreated.

In the third category, false-positive diagnoses are incorrect; patients are not sent away but, instead, are likely to be treated—perhaps needlessly. Treatment may be costly and emotionally and physically traumatic to the patient, and may waste valuable and scarce resources as well. Because in our society the values put upon human life and well-being outweigh economic values, diagnosticians tend to favor positive diagnoses.[1]

The most serious of the four categories is the last. To decide that a patient does not have an illness when he or she does can be a grave, perhaps fatal, mistake.

To each of these possibilities the diagnostician must finally give a yes or no answer, identifying the symptomatic differences between the perceived cause of illness and other illnesses excluded during the diagnostic process. Thus, medical diagnosis follows a yes-no logic that is—or more realistically appears to be—like the two-valued binary logic of computer circuitry.

However, great difficulties attend this simplistic statement. Correct interpretations are often masked by slight differences in symptoms, so that a particular symptom or syndrome can also mean "maybe"—not decisively "yes" or decisively "no" but *probably* "yes," or, to the contrary, *probably* "no." Existence of such uncertainty at any step in the diagnostic sequence compels linkages with other symptoms to reinforce the probability that the outcome will be a true diagnosis.

Fortunately, most diagnoses are true, but even when wrong, they may be harmless. Patients often recover without professional help, sometimes even in spite of it. For such instances the cost of error is likely to be negligible.

There are other instances, however, when uncertainty clouds final diagnoses, and errors, whether false-positives or false-negatives, incur very high costs. Diagnosis for cancer of the cervix in women provides a case in point. Because of the seriousness of the disease, diagnostic procedures are thorough. Most sensitive of the many diagnostic measures is the now-famous Pap test, which has contrib-

---

1. Thomas J. Scheff, "Decision Rules, Types of Error, and Their Consequences in Medical Diagnosis," *Behavioral Science* 8 (1963): 97–107. As part of his thesis that physicians prefer to diagnose illness rather than wellness, Scheff quotes a study of a thousand students who had been examined regarding the advisability of having a tonsillectomy (reported by H. Bakwin in "Pseudocia Pediatrica," *New England Journal of Medicine* 232 [1945]: 693): "Of these, some 611 had their tonsils removed. The remaining 389 were then examined by other physicians, and 174 were selected for tonsillectomy. This left 215 children whose tonsils were apparently normal. Another group of doctors were put to work examining these 215 children, and 99 of them were adjudged in need of tonsillectomy. Still another group of doctors was then employed to examine the remaining children, and nearly one-half were recommended for operation."

Fig. 17.1 Screening curve showing percentage of false-negatives and false-positives for different levels of possibly cancerous cells. Reproduced, by permission, from William J. Horvath, "The Effect of Physician Bias in Medical Diagnosis," *Behavioral Science* 9, no. 4 (1964).

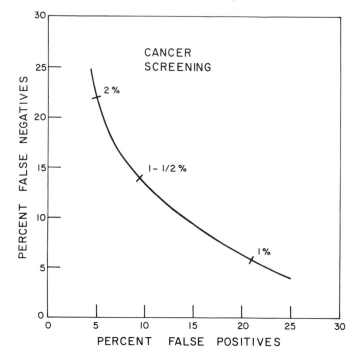

uted significantly to true diagnoses but which cannot be relied upon with absolute certainty.[2] Microscopic examination of smears is used to count suspicious-looking cells. If the number exceeds a chosen small percentage, the test result is considered positive and the patient is very likely to be treated for the disease. If the number is less than the chosen percentage, the result is labeled negative and there may be no treatment. No matter what the chosen threshold of suspicious-looking cells (the yes-no line of demarkation) may be, however, small fractions of false-positives and false-negatives will remain (Fig. 17.1). Either kind of error imposes a heavy penalty: a false-positive tells the patient she has a dread disease when she does not; a false-negative sends the patient away. The cost of a positive error

2. William J. Horvath, "The Effect of Physician Bias in Medical Diagnosis," *Behavioral Science* 9, no. 4 (1964).

is great anxiety and unnecessary and expensive treatment; the cost of a negative error can be the death of the patient.

This example is but one of the many "diagnostic dilemmas" that confront physicians with forced choices between alternatives, choices in which the cost of being wrong is very high.[3] To the great benefit of patients, the diagnostic probabilities for true-positives and true-negatives have markedly improved, with corresponding reductions in the occurrences of false-positives and false-negatives. Increasingly, the computer is being looked upon as a means of diagnostic reinforcement. Two examples of computer-aided medical diagnosis by means of "expert programming" are the subjects of this chapter.

Before narrating these investigations, and to avoid the philosophical passions that attend the subject of artificial intelligence,[4] it seems appropriate to restate Rich's definition, adopted in Chapter 16: *"Artificial Intelligence* (A.I.) is the study of how to make computers do things at which, at the moment, people are better."[5]

Neither of the following examples pretends to have made the computer "think," nor does either claim that the computer is "better than people." Both, however, show that the combination of diagnostic expertise and computer expertise has embodied diagnostic intelligence in the machine. That transfer of intelligence is not yet sufficient to replace human diagnosis by machine diagnosis, but the machine is a useful adjunct that lessens the probabilities of false-positives and false-negatives and thus the penalties they entail. There is also a promise of more and better things to come.

3. M. F. Lechat and Charles D. Flagle, "Allocation of Medical and Associated Resources to the Control of Leprosy," *Acta Hospitalia* 2, no. 2 (1962). These authors discuss a problem analogous to that posed by the Pap test: how many smears should be examined microscopically to diagnose the presence of leprosy in the people of the Belgian Congo? One smear would detect most true-positives and true-negatives, but at a cost of some false-positives and false-negatives. More than one smear would reduce the fraction of false diagnoses but would so strain resources that some part of the population would receive no examination at all.

In an article in J. R. Lawrence, ed., *Operational Research and the Social Sciences* (London: Tavistock Publishers Assoc., 1966), Flagle examines the "diagnostic dilemma" posed by subacute bacterial endocarditis, a disease from which an untreated false-negative can result in death and a false-positive involves six costly weeks of hospital care.

4. See, for example, Frank Rose, "The Black Knight of AI," *Science 85*, March, pp. 46–51.

5. Elaine Rich, *Artificial Intelligence* (New York: McGraw-Hill Book Co., 1983), p. 1.

EMERGENCY ROOM DIAGNOSIS OF PATIENTS WITH ACUTE CHEST PAIN

Patients with acute chest pain often come to hospital emergency rooms and there require prompt diagnosis.[6] About two-thirds of such patients are admitted, many having suffered acute myocardial infarctions, heart disabilities that involve impairment of some part of the muscular substance of the heart. Because of the potentially serious consequences of a false-negative diagnosis, emergency room physicians are encouraged to admit patients if there is diagnostic uncertainty. Uncertainty, however, may result in the admission of patients to coronary-care units, resources that impose high costs upon diagnoses that prove to be false-positives. "If the differentiation between acute myocardial infarction and other causes of chest pain could be made more accurate, the quantity of scarce resources spent on unnecessary admissions to the coronary-care unit could be substantially reduced."[7]

Analysis

To pursue the goal of increasing the number of true diagnoses, the histories of 482 patients who came to the emergency room of Yale-New Haven Hospital between March and December 1977 became the subjects of intensive analysis.

Emergency room interns and residents carefully recorded all diagnostic procedures followed, and did so without knowledge of post-emergency room events; they also recorded their diagnostic and admission decisions. In addition, each "emergency room physican was . . . asked to estimate the probability that the patient's chest pain was attributable to acute myocardial infarction and the probability that the pain was attributable to acute myocardial ischemia."[8] In both instances, the categories chosen were 0 percent,

---

6. This description is based upon a Special Article in the September 2, 1982, issue of the *New England Journal of Medicine*: Lee Goldman et al., "A Computer-derived Protocol to Aid in the Diagnosis of Emergency Room Patients with Acute Chest Pain," pp. 588–96. I am greatly obliged to Dr. Jonathan M. Links, Assistant Professor of Environmental Health Sciences, and Ms. Kathleen Prendergast, Research Project Coordinator, Division of Nuclear Medicine and Radiation Health Science, Johns Hopkins University School of Hygiene and Public Health, for providing me with this example. Neither they nor the authors of the above paper, however, are in any way responsible for errors in my description.

7. Goldman et al., p. 589.

8. Ibid. Ischemic heart disease is caused by blockage that inhibits the flow of

1–5 percent, 6–24 percent, 25–75 percent, 76–94 percent, or above 94 percent.

Follow-up

Follow-up information from 478 of these patients was obtained from six to ten months after visits to the emergency room, to determine whether any chest-pain syndrome had persisted or recurred. Pertinent physician and hospital records also were obtained.

All the patients who had been admitted to the coronary-care unit had been given additional tests. For 11 of these, follow-up information contradicted the hospital discharge diagnosis. For all the others, the hospital discharge diagnoses and follow-up studies were accepted as "ultimate diagnoses."

(The term *ultimate diagnosis* suggests absolute truth, but this may not be so for every case. As said previously, true and false and positive and negative diagnoses can be verified only by subsequent events. Such events may be quick and decisive, as in the case of death and autopsy, but they also may be deferred and uncertain. In this example and the one that follows, "ultimate diagnoses" are "best evidence" standards with which diagnoses by physicians and computer are compared.)

Data from these diagnostic procedures, probability estimates, diagnostic decisions, follow-up tests, ultimate diagnoses, and measures of sensitivity, specificity, and positive predictive value[9] were then subjected to statistical and mathematical analysis. The eventual result was a computer protocol of the kind illustrated in Fig. 17.2. This yes-no decision tree shows 14 terminal branches, designated by the letters *A* through *N*. The first and most discriminating of the questions is that asked at the top, where, as will be shown, the emergency room electrocardiogram (EKG) was most indicative of a positive diagnosis of acute myocardial infarction (MI). Patients' symptoms for which the answer to that first question was "no" were then addressed in the hierarchical order shown in the protocol, with

---

blood to the heart. The disease is serious, but less so than infarction; ischemia may or may not be corrected.

9. *Sensitivity* is the ratio of true-positive diagnoses to the number of patients with the disease. *Specificity* is the ratio of true-negative diagnoses to the number of patients without the disease. *Positive predictive value* is the ratio of the number of true-positive diagnoses to the sum of true-positive and false-positive diagnoses. In each of these ratios, denominators are derived from ultimate diagnoses.

Fig. 17.2 Computer-derived decision tree for classification of patients with acute chest pain. Redrawn from Lee Goldman et al., "A Computer-derived Protocol to Aid in the Diagnosis of Emergency Room Patients with Acute Chest Pain," *New England Journal of Medicine,* September 2, 1982, p. 591.

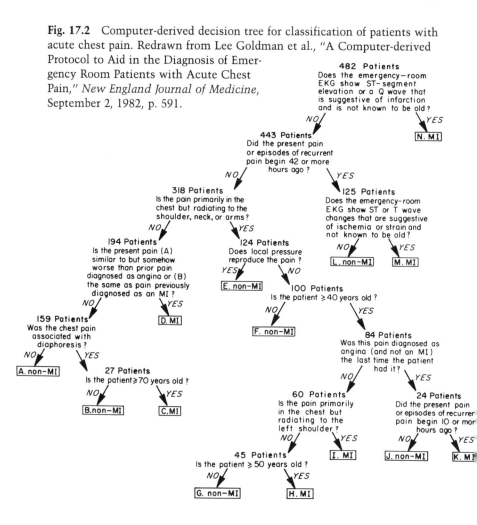

each branching sequence leading to a terminal node indicating either "non-MI" or "MI."

Validation

At a separate hospital, Brigham and Women's Hospital of Boston, two additional sets of patients were made the subjects of similar studies: 357 of these patients were seen in the emergency room between October 1980 and August 1981, and 111 patients who had been admitted in 1978 were subjected to diagnoses by physicians and by the computer program.

Evaluation

Table 17.1 identifies the number of cases of acute myocardial infarction found in each of the 14 terminal subgroups of Fig. 17.2. Of the 145 myocardial infarctions found in the 950 patients in all three sets, the computer correctly identified 143, including all 60 of the infarctions identified in the Yale-New Haven training set.

The computer protocol also performed well in validation testing of the patients sets at Brigham and Women's Hospital, the single exception being a case in subgroup G, where one of the 12 patients in the two groups had an infarction.

The published report describes the evaluation in much more detail than will be given here, my purpose being to compare diagnoses by the expert program with those made by interns and residents.

In general the computer performed about as well as, and in some respects better than, the physicians, and the integrated diagnostic performance of the computer *and* the physicians, taken together, was better than that of either alone (Table 17.2), perhaps because physicians and the computer "missed" different patients.

Despite continuing debate over the precise value of coronary-unit care in decreasing mortality, emphasis in both physician and computer diagnoses was weighted to avoid false-negatives. Because of the seriousness of both myocardial infarctions and myocardial ischemia, this emphasis was well placed, but it did result in the perhaps unnecessary utilization of scarce and expensive coronary-care facilities for some patients. It is possible that diagnostic reinforcement by the computer protocol will contribute to improved cost effectiveness without detriment to patients who do have acute myocardial infarction or acute myocardial ischemia without infarction.

In their report, the investigators concluded that further testing of the protocol would be required before it could be used routinely, and even then, they declared, the protocol would complement, rather than supplant, conventional clinical judgment.[10]

It would be pretentious for a layman to disagree with this modest claim. But while I do not disagree, I cannot forbear observing that both of the participating hospitals have resources that probably exceed, quantitatively and qualitatively, those of smaller, less pres-

10. Goldman et al., p. 595.

Table 17.1 Breakdown of patients in each of the 14 terminal branches of the decision tree of Fig. 17.2

Terminal branch	Yale training set		B. and W. emergency room validation set		B. and W. admission validation set		Total	
	No. of MIs	No. of patients	No. of MIs	No. of Patients	No. of MIs	No. of patients	No. of MIs	No. of patients
A	0	132	2	84	0	13	2	229
B	0	20	0	19	0	9	0	48
C[a]	2	7	1	3	0	1	3	11
D[a]	4	35	4	37	5	12	13	84
E	0	24	0	12	0	1	0	37
F	0	16	0	12	0	1	0	29
G	0	13	1	7	0	5	1	25
H[a]	9	32	4	10	4	16	17	58
I[a]	8	15	1	15	1	6	10	36
J	0	19	1	18	0	8	1	45
K[a]	2	5	1	9	0	3	3	17
L	0	115	1	64	0	6	1	185
M[a]	1	10	4	16	0	6	5	32
N[a]	34	39	35	51	20	24	89	114
Total	60	482	55	357	30	111	145	950

SOURCE: Reproduced, by permission, from Lee Goldman et al., "A Computer-derived Protocol to Aid in the Diagnosis of Emergency Room Patients with Acute Chest Pain," *New England Journal of Medicine*, September 2, 1982, p. 593.

[a] All patients in this terminal branch would be classed as having had acute myocardial infarctions by the computer-derived decision tree.

Table 17.2   Integration of the computer model with the combined emergency room and admission validation sets at Brigham and Women's Hospital

Myocardial infarctions	*Physicians' decisions*	*Computer model integrated with physicians' decisions*
Among patients admitted to coronary-care units	78/241 (32%)	79/186 (42%)
Among patients admitted to other hospital beds	5/48 (10%)	5/107 (5%)
Among patients not admitted	2/179 (1%)	1/175 (0.6%)

SOURCE: Reproduced, by permission, from Goldman et al., p. 594.
NOTE: The data from the computer model are to be compared with actual triage decisions. The integration of the model with the physicians' decisions indicates that a patient would: (1) be admitted to the coronary-care unit (CCU) if the computer model predicted myocardial infarction (MI) *and* the physician recommended hospital admission to the CCU or to another hospital bed; (2) be admitted to another hospital bed if the physician recommended CCU admission but the model did not predict MI, or if the model predicted MI and the physician recommended that the patient be sent home; and (3) be sent home if the model predicted non-MI and the physician did not recommend admission to the CCU.

tigious institutions. If that surmise is correct, a lay opinion may be appropriate here: in hospitals where emergency room resources are meager in either numbers or experience, the protocol could be of great value—and the decision tree could be used at a distance, if need be.

INTERNIST—I

In the preceding example expert programming was applied to a single dominant symptom: acute chest pain, suggestive of myocardial infarction. The diagnostic problem was to distinguish this serious disease from other heart-related possibilities, and to that problem the computer-derived protocol made a valuable contribution.

INTERNIST—I[11] has a much more ambitious objective: to devise a program capable of making multiple and complex diagnoses in internal medicine by forming "an appropriate differential diagnosis in individual 'problem areas' . . . defined as a selected group of ob-

11. Randolph A. Miller, Harry E. Pople, Jr., and Jack D. Myers, "*INTERNIST—I*, An Experimental Computer-based Diagnostic Consultant for General Internal Medicine," *New England Journal of Medicine*, August 19, 1982, pp. 468–76. I am once again obliged to Dr. Jonathan M. Links and Ms. Kathleen Prendergast for this example.

served findings, the differential diagnosis of which is assumed to be a mutually exclusive, closed (i.e., exhaustive) set of diagnoses" (p. 468).

Data base

The knowledge base for INTERNIST–I incorporates individual disease profiles. By inversion of these profiles with computer programs, differential diagnosis can be obtained. The individual disease profiles were to form a hierarchy, organized from the general to the specific, but early experience showed that a rigid hierarchical scheme would be infeasible, because "a single disease often merits simultaneous categorization under more than one heading." By way of example, infectious mononucleosis "is both a hepatocellular infection and a type of infectious lymphadenopathy" (p. 469).

Solution of problems such as these required the development of algorithms that would permit INTERNIST–I to construct problem areas in an ad hoc manner. Three value scales were constructed:

> An interpretation of "evoking strength" to answer the question: "Given a patient with this finding, how strongly should I consider this diagnosis to be its explanation?" (Table 17.3)
>
> An interpretation of "frequency values" as a means of estimating how often patients with the disease have the finding (Table 17.4)
>
> An interpretation of "import values," to express the "global importance of the manifestation—that is, the extent to which one is compelled to explain its presence in any patient" (Table 17.5)

These numbers are judgmental, compiled after review of available knowledge, for "true quantitative information does not exist in the medical literature in most cases" (p. 469).

The knowledge base for INTERNIST–I encompasses over 500 disease profiles and approximately 3,550 manifestations of disease. An example, too lengthy and medically complex to detail here, lists and assigns values for evoking strengths and frequencies to 77 manifestations of alcoholic hepatitis, and follows these with a half-dozen links for that disease.[12] The 3,550 manifestations in the knowledge base are not independent. Properties of each manifestation specify how its presence or absence may influence the presence or absence

---

12. Ibid., p. 471.

Table 17.3 Interpretation of evoking strengths

Evoking strength	Interpretation
0	Nonspecific (manifestation occurs too commonly to be used to construct a differential diagnosis)
1	Diagnosis is a rare or unusual cause of listed manifestation
2	Diagnosis causes a substantial minority of instances of listed manifestation
3	Diagnosis is the most common but not the overwhelming cause of listed manifestation
4	Diagnosis is the overwhelming cause of listed manifestation
5	Listed manifestation is pathognomic for the diagnosis

SOURCE: Reproduced, by permission, from Randolph A. Miller, Harry E. Pople, Jr., and Jack D. Myers, "INTERNIST–I, An Experimental Computer-based Diagnostic Consultant for General Internal Medicine," *New England Journal of Medicine*, August 19, 1982, p. 469.

Table 17.4 Interpretation of frequency values

Frequency value	Interpretation
1	Listed manifestation occurs rarely in the disease
2	Listed manifestation occurs in a substantial minority of cases of the disease
3	Listed manifestation occurs in roughly half the cases
4	Listed manifestation occurs in the substantial majority of cases
5	Listed manifestation occurs in essentially all cases (i.e., it is a prerequisite for the diagnosis)

SOURCE: Reproduced, by permission, from Miller, Pople, and Myers, p. 470.

of other manifestations. About 6,500 such relationships are included in the knowledge base.

Procedure

As Tables 17.3, 17.4, and 17.5 indicate, diagnoses depend upon probabilistic estimates, but the behavior of INTERNIST–I derives "primarily from the application of two heuristic principles: the formation of problem areas through a partitioning algorithm and the conclusion

17.5 Interpretation of import values

Import value	Interpretation
1	Manifestation is usually unimportant, occurs commonly in normal persons, and is easily disregarded
2	Manifestation may be important but can often be ignored; context is important
3	Manifestation if of medium importance but may be an unreliable indicator of any specific disease
4	Manifestation is highly important and can only rarely be disregarded, as, for example, a false-positive result
5	Manifestation absolutely must be explained by one of the final diagnoses

SOURCE: Reproduced, by permission, from Miller, Pople, and Myers, p. 470.

of diagnoses within problem areas, using strategies such as diagnosis by exclusion" (p. 475).

As many as 11 steps are taken during an INTERNIST–I diagnostic consultation. To avoid medical complexities and for the sake of brevity, six of these steps are condensed below (at some sacrifice of clarity) and five have been omitted:

1. Initial positive (present) and negative (absent) findings in the patient are entered by the user. As each new positive manifestation is encountered, the program retrieves its complete differential diagnosis from the inverted disease profiles in the knowledge base. A "disease hypothesis" is created for each item on the manifestation's differential diagnosis list. . . .

2. For each disease hypothesis, four lists are maintained: all positive manifestations . . . that are explained by the disease hypothesis . . . ; all manifestations that might occur in a patient with the disease but are known to be absent in the patient being considered; . . . all manifestations that are present . . . but not explained by the disease hypothesis; . . . and manifestations that are on the disease profile but about which nothing is known. . . .

3. Each hypothesis on the master list of diagnoses is given a score. . . . The positive component includes the weights of all manifestations explained by the hypothesis. [For evoking strengths a nonlinear weighting scheme is used: for an evoking strength of 0, 1 point is assigned; a strength of 1 counts as 4, 2 counts as 10, 3 counts as 20, 4 as 40, and 5 as 80.] . . . Any disease hypothesis related to a previously concluded diagnosis . . . is given a bonus score. . . .

The negative component includes the weight of all manifestations that are expected to occur in patients with the disease but are absent in the patient under consideration. . . . [This scale also is nonlinear: a frequency of 1 counts as -1 point, 2 as -4, 3 as -7, 4 as -15, and 5 as -30.]

. . . Also included are the weights of all manifestations present in the patient but not explained by the hypothesized diagnosis. The import (clinical importance) of each manifestation is used to assess this penalty. [An] import of 1 . . . [counts as −2, a 2 as −6, a 3 as −10, a 4 as −20, and a 5 as −40]. The net score for any disease hypothesis is thus the sum of the four component weights.

4. After all disease hypotheses have been scored, the master list of all hypotheses is sorted by descending scores . . . [and] scores [that] fall a threshold number of points below the topmost diagnosis are temporarily discarded as unattractive.

5. At this point, the sorted master differential diagnosis list is a heterogeneous grouping of many disease hypotheses. A critical step in the diagnostic logic . . . is to delineate a set of competitors for the topmost diagnosis. . . . Only one of a set of diseases in a properly defined problem area is likely to be present in a patient.

6. Once the problem area containing the most attractive diagnosis has been selected, criteria for establishing a definitive diagnosis can be applied. If the problem area contains only the topmost diagnosis, INTERNIST–I will immediately decide on (conclude) that diagnosis. (p. 475)

The five omitted steps delineate additional diagnostic details covering procedures to be followed when outcomes are less definitive; the manner of questioning; recycling the program; deferral of diagnosis when "all useful lines of questioning have been exhausted"; and stopping the program. An appendix that occupies more than three columns of closely spaced small type—much too long and complex to be included here—gives a sample case analysis that lists 50 "initial positive manifestations" and 15 "initial negative findings" and concludes with a diagnosis of hepatic encephalopathy.[13]

Evaluation

Nineteen clinically challenging cases that had been the subjects of "clinicopathological conferences" (CPCs) and that had been published in the *New England Journal of Medicine* were chosen in order to compare the diagnoses made by INTERNIST–I with those made by clinicians at the Massachusetts General Hospital. As in the case of the protocol evaluated in the myocardial infarction study, CPC diagnoses were analogously "ultimate."

13. Ibid., p. 476. Arbitrarily chosen from among the 65 items are the following positive and negative manifestations: + Age greater than fifty-five; + Depression HX; + Jaundice; + Liver enlarged moderate; − Diarrhea chronic; − Fever; − Uric acid blood increased.

Table 17.6 Summary of results for major diagnoses in 19 cases used in the INTERNIST–I evaluation

Category	INTERNIST–I	Clinicians	Discussants
Total number of possible correct diagnoses	43	43	43
	Number of instances		
Definitive, correct	17	23	29
Tentative, correct	8	5	6
Failed to make correct diagnosis	18	15	8
Definitive, incorrect	5	8	11
Tentative, incorrect	6	5	2
Total number of incorrect diagnoses	11	13	13
Total number of errors in diagnosis	29	28	21

SOURCE: Reproduced, by permission, from Miller, Pople, and Myers, p. 473.

Given the difficulty of the CPC cases and the ad hoc nature of the computer's algorithms, INTERNIST–I "performed remarkably well," but it has not yet been proposed for clinical application.

Table 17.6 shows the comparisons. For the 19 cases analyzed, there were 43 "possible correct major diagnoses," numbers that suggest the complexity of the cases chosen. INTERNIST–I, the clinicians, and the case discussants respectively made 17, 23, and 29 "correct definitive diagnoses." In this same order there were 8, 5, and 6 correct tentative diagnoses, leaving, from the 43 possible diagnoses, 18, 15, and 8 "misses." The total numbers of incorrect diagnoses, definitive and tentative, were respectively 11 for INTERNIST–I, 13 for the clinicians, and 13 for the CPC discussants. INTERNIST–I made the most errors (29), the clinicians made 28, and the discussants, 21.

While emphasizing the power of their underlying heuristic methods in the field of internal medicine, the authors of the Massachusetts General Hospital study were candid in their evaluations of INTERNIST–I's shortcomings:

INTERNIST–I's greatest failing . . . was its inability to attribute findings to their proper causes. . . . the program cannot synthesize a general overview in complex multisystem problems. . . .

. . . To its detriment, INTERNIST–I's handling of explanation is shallow. When the program concludes a diagnosis, that diagnosis is allowed to explain any observed manifestations that are listed on its disease profile. Once explained, a manifestation is no longer used to evoke new disease hypotheses or to participate in the scoring process. . . .

. . . But INTERNIST–I cannot formulate a broad perspective in complicated multisystem problems. It is constrained to working with tunnel vision, discriminating among diagnoses within each problem area and remaining unable to look at several problem areas simultaneously. . . . New programming approaches to complex reasoning processes have been developed to enable CADUCEUS, the successor to INTERNIST–I, to synthesize a broad overview incorporating causal relations into an approach to a patient's problems. (p. 474)

These criticisms, excerpted from among others, reveal commendable objectivity and professional spirit. Emphasis upon deficiencies seems proper, and so do modest predictions about better things to come. The declaration that INTERNIST–I is not yet ready for clinical application shows restraint and wisdom.

OPINION

To one who is not expert about computers or medicine, these remarkable investigations of exceedingly complex problems are too modest: the potential for computer-aided diagnoses has not been sufficiently claimed. Earlier, in evaluating the protocol for diagnosis of acute chest pain, I pointed out that relatively few hospital emergency rooms have the rich resources of Yale–New Haven, Brigham and Women's, or Massachusetts General. For less well endowed establishments and for the patients who use them, the computer-derived protocol could be a clinical boon. That opinion applies to INTERNIST–I as well. Beyond that, the knowledge base so painstakingly and laboriously created in each case constitutes a compendium of symptoms that appears to be correctable, expandable, and amendable to computer search.

The human diagnostician's knowledge base is derived from rigorous training and experience; its clinical application depends upon tests, recall, and heuristic search, reinforced by abilities the computer does not yet have: intuition, cognition, perception, feeling.

Human recall, however, is not totally dependable, even when augmented by consultations. Human memory lapses are not uncommon, and search sometimes fails to reveal what may be critical lacunae. Given the power and speed of computers in performing both comprehensive and heuristic search, is it not reasonable to believe that computer-*aided* diagnoses may even now be better than diagnoses made by humans alone?

Both of the examples discussed in this chapter serve to emphasize an important point: for computer-aided systems the efficiency ratio of output to input probably has been reduced—for example, by the augmentation of the diagnostic process with the computer. But considering the human cost of possible error, efficiency ratios have little meaning. The criterion becomes *effectiveness*—the utility of being "right." As computer-aided diagnostic "systems" make better diagnoses possible, they will become even more effective.

Both examples show that human intelligence has been organized and embodied in the machine. Whether these achievements of expert programming will be regarded as artificial intelligence I do not know. I leave that question to the hair-splitting syllogisms of philosophers.

EIGHTEEN

Conclusions

As defined in this book, an operational system requires a succession of related operations that provide goods or services to customers and clients. The manner in which these related operations are arranged and performed determines the system technology.

Most operational systems are composed of combinations of different kinds of arrangements: they are, using an evolutionary term, *hybrids*. The characteristics of these components are identifiable and definable, as if each existed in "pure" form. Like "perfect gases," "perfect elasticity," and "perfect competition," "perfect system technology" is a useful fiction.

Seven of these "pure" operations technologies have been identified, defined, and exemplified. They are:

Jobbing systems
Articulated systems
Balanced systems
Continuous systems
Automated systems
Programmed systems
"Intelligent" systems

This list depicts an approximately chronological, overlapping sequence. Although there is no biological evidence to support the supposition, there appears to be an analogy, so in fancy the sequence has been portrayed as evolutionary, with each system technology, by a kind of mutation, emerging as a species or subspecies of its forebears.

The evolutionary progression of operational system technologies has been marked by the increasing dominance of mechanisms that are now to an increasing extent controlled by computer infor-

233

mation. For a time, technological progression involved sacrifices of diverse capabilities—variety—in favor of speed; now capability for variety has been restored, without loss of speed.

Time is the dimension by which the seven system technologies discussed here have been compared. Processing times, in-process delay intervals, moving times between sequential work centers, sequencing delays, and systems stoppages have been used to postulate equalities and proportionalities for passage of work through each kind of "perfect" technology.

Total time and cycle time, respectively, are summations for passage of work through a system and for passage through the system's shortest path. Operational efficiency is the familiar but slightly modified output/input ratio. System efficiency, important but seldom measured, also is a ratio, with in-process delays and stoppages added to the denominator. Proportionalities, expressed in units of time, have been modeled for working-capital requirements, in-process inventories, and operational floor-space requirements.

Substitution of numbers in these models for the most part has been conjectural but is reasonably typical of the system technologies to which the numbers have been applied. As mutations in system technologies have occurred, the time intervals symbolized have undergone dramatic changes in significance. Processing intervals have become smaller and smaller, approaching zero in present-day computers. In-process delays and moving intervals, and sequencing delays as well, have come to have much less meaning, while stoppages have become much more important.

Stoppages and the capital and operating costs associated with their mitigation, and the very large costs deriving from stoppage durations (the years-long stoppage of Unit 1 at the Three Mile Island nuclear power plant comes to mind as an extreme example), have proved impossible to quantify; they have been discussed, but for the more recent technological mutations, the time-dependent models lack arithmetic verisimilitude. Nevertheless, plausibility is claimed for the system comparisons made.

Equally immeasurable has been the quality of system *effectiveness*, the degree to which any system fulfills its stated mission to the satisfaction of those providing and those receiving the system's goods and services. While there probably is a direct relationship between system efficiency and system effectiveness as a general rule, there are exceptions: it is possible for a system to be egregiously inefficient and, at the same time, very effective; it is also possible for a system to be efficient and ineffective; and it is possible to change either without affecting the other. Despite the subjective

character of effectiveness, judgments have been made about perceived changes when it has seemed proper to do so.[1]

JOBBING SYSTEMS

Progenitor of all operating system technologies, jobbing systems are
designed, equipped, and staffed to meet the diverse needs of customers and clients. Capability for variety is the great advantage enjoyed
by jobbing systems, along with relatively lower fixed-capital requirements, specialization to meet the demands of particular yet
broad and stable markets, and possible growth by gradual accretion.
    Offsetting these advantages, however, are serious disadvantages:

> Variable work-center demands and variable work-center se-
> quences create load instabilities that almost always are met
> by in-process inventories that create queues of work
> preceding and following actual processing. Very long delays
> attend every work center, and these accumulate for every
> unit processed and at every work center.

> Moves from work center to work center also vary in distance
> and time and must be initiated and directed lest work be
> lost or go astray.

> Unless there is a first come–first served rule at each work
> center—which is almost never practicable—control of order
> sequences must be exercised at every work center. Too often
> this is not practicable and must give way to control of key
> work centers. Control of work in jobbing systems all too
> frequently is chaotic.

These and other disutilities of jobbing systems result in long intervals for total times and cycle times, low operational and very low
stem efficiencies, and large inventories of work in process that demand substantial working capital and ample floor space.
    Division of labor, the creation of multiple work centers and
work stations, exacts little penalty in later system technologies, but
it exacts a high cost in the before- and after-processing delays at
jobbing system work centers. Reducing division of labor by combining compatible work centers or, better still, eliminating those
found to be unnecessary, offers a very practical way to improve

---

1. For an example see Robert H. Roy, *The Administrative Process* (Baltimore:
Johns Hopkins Press, 1958), chap. 5, §4, and p. 46.

jobbing system performance—that is, system effectiveness as well as system efficiency.

Deficiencies in control can also be remedied by adoption of established means of forecasting and reporting and, more simply, by what have been called $t + \Delta t$ limits upon events. Examples of both have been given.

Despite jobbing system deficiencies, it is possible, perhaps probable, that the number of jobbing systems has increased throughout the operational world. Evidence for this assumption is provided by the increase that has occurred in service employment, in organizations that are most likely to operate in the jobbing mode. Government bureaucracies at the local, state, and national level typify the situation: they are characterized by extensive division of labor, little control of operations at multiple work centers, and consequently long in-process delays.

Improvements, very large improvements, in system efficiency and effectiveness could be realized by the means cited. Given the inertial forces of bureaucracy and the behavioral changes that would be necessary, however, these desirable outcomes are most unlikely.

ARTICULATED SYSTEMS

An articulated system has been defined as an arrangement of work centers and component work stations in fixed and approximately balanced sequence, designed to provide but one kind of product or service.

By sacrificing capability for variety, articulated system technology assumes a risk of commensalism—dependence upon a continuing demand for a single product or service—but gains in ways sufficiently dramatic to warrant calling the articulated system a "punctuational" evolutionary mutation:

> Control of the flow of work through the system is embodied in the system's "prenatal" design. It is necessary only to start work at the first center to have it proceed very rapidly through the fixed sequence to completion.

> Fixed sequences of operations reduce moving times and distances, and uniformity in product or service eliminates changeovers and makes possible special-purpose equipment and refined methods. Process intervals are shortened, to the betterment of operational efficiency.

Approximate balance yields very small pre- and post-operation delays, and correspondingly smaller in-process inventories, working-capital requirements, and in-process floor space. Total time and cycle time are dramatically shortened.

By these eliminations and reductions system efficiency is much improved, as is service to customers and clients. In general, articulated systems are more effective than jobbing systems.

These advantages of articulation are to a degree offset by associated necessities:

To the extent that there are imbalances, there are likely to be sequencing delays.

Achieving and sustaining balance demands employee and operational flexibility.

There being only small in-process inventories, stoppages can be very costly. To guard against them, stand-by equipment and personnel can be provided, at extra capital and operating costs.

For articulated systems (and increasingly for balanced, continuous, and automated systems), logistical services must be synchronized with flow of work; provision of raw materials, supplies, and tools becomes a system necessity.

Meeting these needs diminishes the attraction of articulation, but not by much; the margin of superiority remains large. For large and stable markets, the exchange of sameness and speed for variety has provided benefits to those who operate articulated systems and those who are served by them.

BALANCED SYSTEMS

Balanced systems may be characterized as mutants of articulated systems; epitomized by the balanced assembly lines introduced by Henry Ford,[2] they are a kind of evolutionary refinement.

In balanced systems, processing intervals at each work station,

2. Allan Nevins and Frank E. Hill, *Ford, The Times, the Man, the Company* (New York: Arno Press, 1976), vol. 1, chap. 18.

in the "perfect" sense of a "pure" system, are exactly, not approximately, equal. Pre- and postoperation delays become zero, there are no sequencing delays, and work stations are juxtaposed so that moving times and distances are very short.

The yield from balanced systems operating in this fashion is considerable: total times and cycle times are diminished still more, operational and system efficiencies rise, and requirements for working capital, in-process inventory, and operational floor space are further reduced.

This degree of "perfection," however, is difficult to achieve and sustain. Most balanced system technologies are found in assembly operations run by humans, each of whom must process in concert. Subdividing the work with such exactitude is difficult at the design stage; so also is effecting balance during the period of shakedown. Human endeavor and aptitude differ from one worker to another, and humans, unlike machines, suffer from boredom and fatigue. To a degree, balance can be reinforced by machine pacing, which has given rise to many labor-management conflicts and widespread concern about the stultifying effects of assembly lines.

As in other advanced operational system technologies, stoppages are costly for balanced systems. Let one operator fall behind and others must follow suit; let one stop and all must do so. To guard against such delays, redundant, broadly skilled "roving" operators are often employed to step in at any bottleneck to help that operator regain synchrony with work flow. Like other kinds of redundancy, this measure entails cost that in some degree reduces balanced system technological advantage.

CONTINUOUS SYSTEMS

The contribution of continuous system operational technology to the evolutionary sequence has been to process work while it is in motion. In this sense, movement itself becomes a work center; thus, in effect, any "pure" continuous system becomes one single work center.

Processing during movement requires that there be either a stream of product or discrete units so closely spaced as to resemble a stream.

Continuous systems are capital intensive and require instrumental, rather than human, controls. In such systems humans are likely to become monitors rather than operators, enduring monotony while all goes well, yet possessing sophisticated knowledge to prevent or rectify costly stoppages.

Instrumental controls, important in continuous systems, have had enduring significance, in extending and refining the use of feedback by extracting a small fraction of a system's energy to control the system itself.

AUTOMATED SYSTEMS

"Perfect" automated systems are devoid of human participation and are, except for services of energy, supply, and maintenance, totally capital intensive. No such perfectly automated systems exist, but parts of systems, to a greater and greater extent, are being made to operate automatically, to process work, move it to a successor work station, process it there, move it once again, and, with continued process and move balance, deliver the finished product from the system's final stage.

Moving, positioning, processing, halting, and again and again repeating these actions requires signals to initiate, operate, and stop each successive step. These signals may be given by mechanical, electrical, hydraulic, thermal, or any other means of sensing a difference. All of them depend upon the feedback of fractions of the system's energy for purposes of control. Servomechanisms are a common means of effecting automated controls, but it is more pertinent to rely upon current vernacular here: automated systems are controlled by "hardware," in contrast to the "software" controls of programmed systems.

It has often been expensive to alter automated systems. A transfer machine can automatically perform dozens of repetitive operations but it must be replaced if the product is redesigned.

Two developments associated with automated system technology have alleviated the need for and cost of such retooling: the introduction of analogs, models that can be followed by a stylus to guide a tool; and the use of "numerical controls," which achieve the same result. Operational changes can be accomplished by changing the analog or by changing the numbers that constitute the control signals.

These developments have been the first steps in closing the metaphorical circle: they have introduced low-cost means for restoring system capability for variety without sacrifice of speed.

PROGRAMMED SYSTEMS

The programs by which computers are instructed—"software" as compared to the "hardware" of automated systems—have provided the capstone for restoring variety to the capabilities of operational systems. Retooling, so often a costly necessity in articulated, balanced, continuous, and automated systems, can, in programmed systems, be accomplished by simply changing programs.

Regaining system flexibility without sacrificing speed has not always been easy, but the impact of programming upon the operational world has been revolutionary, a punctuational step in the evolution of operations technologies.

Each of the predecessors of programmed systems has benefited from programming technology: Jobbing systems in banking, telephony, libraries, transportation, and government have become programmed technologies. Articulated, balanced, continuous, and automated systems of the kinds that have been discussed are now controlled and monitored by computers and the programs by which they are instructed. We are in the midst of transitions that undoubtedly will proliferate as more and more of man's intelligence is embodied in computers and their programs.

"INTELLIGENT" SYSTEMS

As noted earlier, the word *intelligence* defies precise definition, and when combined with the adjective *artificial*, it evokes passionate controversy. To some, artificial intelligence connotes an unlikely someday computer programmed to embody the whole range of man's senses and capabilities. To others, artificial intelligence has a less ambitious goal. In the words of Elaine Rich, *"Artificial Intelligence is the study of how to make computers do things at which, at the moment, people are better."*[3]

The latter, more feasible definition appeals to my sense of order and seems also to fit the evolutionary metaphor I have proposed. Since the beginning of tool making, more and more of man's motor abilities, his skill and dexterity, have been transferred to, embodied in, machines. Gradually, elements of man's mental skills also have been transferred to machines. Computers and their programs have

---

3. Elaine Rich, *Artificial Intelligence* (New York: McGraw-Hill Book Co., 1983), p. 1.

given giant impetus to these transfers of mental ability, here equated with "intelligence."

Engaging in making computers "do things at which, at the moment, people are better" goes on apace, in ways that are at once arcane and fruitful. There are new algorithms, new programming languages, new and more powerful computer architecture, and new and useful applications—and many more will follow.

Let me conclude this summary with a few predictions, some of which have already been made:

Jobbing systems will continue to abound in many organizations, especially those in government. It is unlikely that many will improve their performance—either their efficiency or their effectiveness—by adopting meaningful controls or reducing division of labor.

As has already happened, however, some jobbing systems will be converted to programming technologies.

Articulated, balanced, continuous, and automated systems will continue to perform large amounts of the world's work, but increasingly they will embrace programmed controls, thereby enhancing their capabilities to achieve variety.

Programmed operational systems will continue to proliferate and inevitably will embody more and more of man's mental and motor capabilities.

Index

Robert H. Roy is Dean Emeritus of Engineering and Professor Emeritus of Industrial Engineering at the Johns Hopkins University. Two of his previous books are available from Johns Hopkins: *The Administrative Process* and *The Cultures of Management*. The latter received the Book-of-the-Year Award from the Institute of Industrial Engineering.